Health Care in Transition

Immunizing the Human Body

The Role of Non-Descript and Traditional Edible Green Leaves

HEALTH CARE IN TRANSITION

Additional books and e-books in this series can be found on Nova's website under the Series tab.

HEALTH CARE IN TRANSITION

IMMUNIZING THE HUMAN BODY

THE ROLE OF NON-DESCRIPT AND TRADITIONAL EDIBLE GREEN LEAVES

A B SHARANGI
S K ACHARYA
AND
SUDDHASUCHI DAS

nova
Medicine & Health
New York

NOTICE TO THE READER

Library of Congress Cataloging-in-Publication Data

ISBN: 978-1-53619-136-3

Published by Nova Science Publishers, Inc. † New York

CONTENTS

FOREWORD

M. S. Swaminathan

The whole planet is reeling under Covid-19 pandemic, lives approaching close to two million have already been claimed by this 'apocalyptic threat', but we are moving with a stern belief that both science and humanity shall overcome! Beyond the flashing of millions of news-cameras, the farmers have not quitted the farm, silently with resolute determination, they have kept the civilization assured of food supply. Similarly, beyond the fatigued attention and cliches of conventional research, the nondescript green leaves and blades have been with their vigor and euphoria to serve the human civilization for more than thousands years of their existence. This is the time to ponder and act to accentuate the paradigm shift from overexploited and depleted gene resource to relatively unattended gene pool in more than 6 lakh rural villages in India including peripheral urban settlements harboring on huge diverse nondescript biodiversity. When any ecology remained rather unintervened by humane need and greed, it has its own glory of ecological resilience and elasticity. Chemical fertilizers or pesticides, when administered beyond proportion, the ecological service is certain to be jeopardized in favour of chaos and dissonance. Besides, edible plants grown in an uninterfered ecology, it retains immense food values and natural potency of generating immunological enabling in human physiology.

I hope this book, extremely innovative in nature, shall add scholastic aroma to the researchers and policy makers.

M. S. Swaminathan

PREFACE

Land is the mother of civilization, agriculture is the mentor. Both are having a descript and nondescript statuses; one is, under mass cognition, when other goes incognito. The all encompassing aggression of urbanite lifestyles and settlement, the land becomes increasingly scarce and dearer, making availability of food to each hungry bowl very difficult and complex. The nondescript land and agriculture including horticulture and livestock are offering us an opportunity for a convincing and graceful survival amidst impending food crisis and malnutrition, even more, amidst a holocaust of COVID-19 pandemic. Nondescript leafy vegetables are the treasures to humanity, because these are not yet wrongly handled by the need of claustrophobic technocracy of so called modern agriculture. Being dwelt in a Holon of natural ecosystem, these can provide the panacea to much needed immunity building, nutrition and green healthcare to millions on this planet. Hope, this book will make a breakthrough in a non-descript pathway in walk of a new age agricultural and ecological science beyond the fences of discipline and geo-spatial confinement.

A. B. Sharangi
S. K. Acharya
S. Das

Chapter 1

NON DESCRIPT LAND: SCOPE OF EDIBLE GREEN LEAF PRODUCTION FOR IMMUNITY

A nondescript mixture of styles in the worst possible taste....
George Bernard Shaw

ABSTRACT

The non-descript approach to agriculture is emerging as an important and inevitable option to achieve ecological resilience and sustainable production for the humanity. As an aftermath of chemical intensive and machine driven agriculture, the resilience of the operating agro-ecosystem has been disrupted seriously. In the name of intensive-integrated farming, we have disintegrated water from soil, soil from biodiversity and all from humanity. This is the high time to go for non-descript pathway by re-integrating knowledge with ecology, ecology with pristine, overlooked non-descript entities like folk, green vegetables growing in an ecosystem peacefully unattended by modern agriculture. We need to make ecology restoring resilience, resources are to both extrapolate and interpolate with non-descript ecological subjects dancing with the pride of nature since time immemorial.

Keywords: Ecology, resilience, non-descript, biodiversity, green vegetables

The lexical stands out of the term 'non-descript' means belonging or appearing to no particular class or kind: not easily described. The era of modern agriculture has been conspicuous by two distinct trends, one is, increasing reliance on high yielding seed, fertilizer, pesticides and ground water extraction for ushering green revolution; on the other hand, a pattern of persistent negligence to traditional knowledge, wisdom and non-descript aspect of natural agriculture. This has invited an isochronous problem of environmental degradation and a social divide there after, those who can access the benefits of modern technology and those who cannot. So far we kept on relying on only 25-30 species for producing almost sixty per cent of global food requirement. The gene myopia in agriculture is one of the reasons why food genes are so fatigued, these are over exploited either. Not only some genes supporting our food security are fatigued, the whole of agro-ecosystem is heavily exploited and extorted as well. Our ecosystem has been fragile in both vertical and horizontal dimensions. The huge dependence of ground water reserve for wheat and summer rice cultivation causes serious plunge of ground water table and invites heavy contamination of toxic metallic ions like arsenic.

The monolithic exploitation of nutrient within 1-6 cm soil regime by some selected variety with similar root morphology has damaged the soil health, which has now been enough fragile to add a stubborn stagnation of total food grain production for the last two decades in a situation where global population is with unabated growth and is well destined to reach 9-9.5 billion count. It speaks alarmingly high when we need to enhance at least 60 per cent food production over its present level.

The issues of poverty remain ever simmering and with an increasing aggression, it now starts engulfing South East Asian countries. Only India has become the dwelling house of 370 million people suffering from extreme poverty. If 290 million tonnes of food grains produced by India in a year is really a grand success, then why should 270 million having remained hungry over the years! In addition, despite undeniable progress in reducing rates of undernourishment and improving levels of nutrition and health, almost 800 million people are chronically hungry and 2 billion suffer micronutrient deficiencies (FAO, 2012). Poverty has further been rudely humiliated by poverty, malnutrition and imbecile immunity, of course denial to the class discrimination to include other echelons of economy in society. That is how, the COVID-19 pandemic has been able to execute the carnage of human lives with a count approaching to two millions.

Why Non-Descript Concept Figures up and, Is It the Glory of an Un-Intervened Ecology?

The non-descript biota are being maintained in an un-descript ecosystem, which are reigning far away from human attention. These are marshy lands, bogs, lagoons, barren lands, water bodies by the side of railway tracks or highways, even the roof tops. Here ecology has her own scripts of resilience, which can dance with a jovial resilience. By remaining pristine from heavy metal loads from chemicalized agriculture or nitrate toxicity, they are nurturing flora and fauna as if a mother is offering her nectar to her new born baby. When ecological resilience is naturally maintained, the vegetables or even fishes are supposed to undergo an uninterrupted physiological growth and reproductive exuberance. The plant tissues are being loaded with minerals, vitamins, anti-oxidants and bio-safety ingredients far more than anything grown in a corrupt and modern farm. Non-descript lands and biota are the placid debonair scripted by nature in a captivity to remain pure and pristine beyond greedy eyes of self-centered consumerisms of so called humanity.

The Aggression of Market Driven Modern Agriculture and Ruthless Depletion of Eco-Resources

All over the world, the role of modernizing agriculture has been either to feed hungry millions or to defend industry and urbanization. The pressure or provocation or both have been so intense that turns precarious in the form of mindless exploitation of water, soil, biodiversity and denial to traditional wisdom. When famine is defended, commerce keeps soaring up. The myopic approach to gene diversity, low risk-high gain technology and shorter duration with higher profitability have brought into the core management of crop productivity. Tons of pesticide including synthetic pyrethroids started killing target and non-target insects to add more taxation to a fragile ecosystem. We started forgetting them if pollinators are killed as insect pest, the ecological resilience would be seriously affected. When a centimeter of soil layer demands few hundred years to create, soil erosion was aggravated due to intense and wrong farm mechanization.

Non-Descript Agriculture: Is It a Nemesis to Follow Either?

Not exactly, rather, in another way round, a changed and re-engineered or redeemed approach to invite and involve overlooked and set aside realms of agro-ecological perspectives of operating agriculture. If traditional pieces of land left rather unused or leafy vegetables that grow in unattended marshy lands need to be brought back to diversify existing ecological functions. It would add more immunological properties along with huge elements of therapeutics. It will save conventional land mass, add redefined land acreages and redeemed water resources. So, we need to have a policy change as well as paradigm shift. When soil is converted to square meters of concrete and reinforced roof top, why don't we rename the same rooftop as non-descript land, to be splendidly converted into a garden or vertical farming. Whenever tens and thousands hectares of roadside water bodies are restored from becoming a sink for eutrofication into a sanctuary for local icthyiofaunal diversity along with an aqua production site for ipomoea vegetables, it adds a belligerence to the performing eco-system with generation of sustainable livelihood as well.

The need of the hour is to trigger a process of changing attitude and a reconfigured technology options for those we need it the most or the worst.

The Classification of Non Descript Land

Across the globe, the expansion of land under cultivation is becoming extremely inelastic and with the unabated expansion of urban habitats, the availability of land is squeezing very fast. So this is the imperative of the situation to go for non-descript lands, the lands which are not otherwise in use or under conventional agriculture. The lands are sometimes in taxonomical classification of barren or waste lands, grasslands or bogs. Here is a classification of nondescript lands:

i. Current Fallow
ii. Fallow Land other than Current Fallow
iii. Cultivable Waste
iv. Misc. Tree Crops & Groves
v. Permanent Pasture & Other Grazing Land
vi. Land put to Non-agricultural Users
vii. Forest

The Potential Crops for Edible Green Leaf Production

Many common vegetables can be grown in non-descript land areas for edible green leaf production. Besides, some uncommon vegetable crops like broccoli, Chinese cabbage, asparagus, celery, parsley, leek, baby corn, cherry tomato, etc, can also be tried with suitable management practices. A study conducted by Suman (2008) revealed that only 13 per cent farmers produced the seeds of pea, French bean, palak, okra for own use for next season. 16.67 per cent farmers were satisfied from the income of vegetable production. Spices are emerging crops acclaimed not only in India, but also throughout the world. However, to promote the seed requirement by the farmers due care is to be taken in West Bengal areas for edible leaf production of several spices. Our world is endowed with a vast area of land having differential land and agro-climatic situations. To grab this unique opportunity several plantation crops can be grown particularly to supplement the huge requirement of seed and quality planting materials to the growers. These plantation crops include coconut, arecanut, cashewnut, betelvine, etc. Several other crops like flowers and ornamentals, medicinal and aromatic crops, forest tree crops etc can also be grown in the non-descript land situations for edible green leaf production.

Figure 1.1. The paradigmatic Structure of non-descript land (NDL).

Table 1.1. Some typologies of Non-Descript Lands (NDL) with local name, land type and features

Sl No	Local name	Land type	Features
1	Nayanjuli	Marginal aquatic body by the roadside bunds	Perennial to seasonal water Availability of local fish species Water weeds, snails and water chestnuts.
2	Smashan (Burning Ghat)	Wasteland generally associated with watering.	Generally raised land with local trees, herbs, shrubs, weeds, away from community settelment.
3	Traditional and cultural waterbodies E.g., Dighis and Bunds	Traditional waterbodies with raised embarkments with huge floral and faunal diversity	Local icthyofaunal diversity, minor fruits, trees, rare type of waterweeds and birds.
4	Bhusundir Math	Abandoned lands with beleifs, taboos and far-away watching	Occasionally tried with field crops, local pulses and social forestry, rich with NPK status and traditional biotypes.
5	Open roof tops: Khola Chhad	Open terrace 10-100ft above ground level, closer to canopy meteorology 3-4 lakhs ha of open terrace in \West Bengal transformed landmass.	Domestic grooming on movable roots, plastic house with rain water harvesting or municipal water supply.
6	Basra	A piece of land not used for cultivation and used only for keeping waste materials of farm	Post harvest operations for value addition, extracting marketable products and spending recess hours
7	Bali khadan	A piece of land or stretch of land used for extraction of sands for construction and other civil works	Commercial extraction of sands, destruction of natural topography, livelihood generation
8	Eeet bhata (Brick kiln)	The land area where the bricks are manufactured	Conversion of natural lands into hard and non-recyclable masses
9	Char	The area emerged from the river	Rich alluvial soil, remains fertile for upto 6-8 months

The unique opportunity and benefits of non-descript lands:

- Economic viability may be impaired by socio-cultural issues.

- Around these lands, there a lot of taboo, prejudices and ritual barrier to invite modern technology in this piece of land.
- Policy barrier: A bane to adopt the land for conventional Agriculture.
- Technology Non-availability:
- Need based research vs. Subjective research
- Alternative Technology vs. General Technology

The Problems Ahead and Strategies to Overcome: The Typical Problems in Dealing with Non-Descript Land Resources

1) Entitlement is a problem: Ownership to this land is mostly community-based, sometimes private, and other times it goes controversial.
2) Research inputs are very poor on the land utilization technology for this ND lands.
3) The quality of the lands in terms of NPK status, run-off character, marketability, participatory micro planning is not attempted yet.
4) Social access to the economic production is not customized.
5) Lands are sometimes very difficult to access.
6) Beliefs and taboos related with these pieces of land may create some hindrance towards opting for economic activities.
7) Absence of proper State Policy is one of the major problems.

Peri-Urban Social Ecosystem: An Emerging Trend across the World

Of late, peri-urban agriculture is gaining attention for its potential role in increasing food security, employment and income generation, poverty alleviation, community resource development, waste management and environmental sustainability by utilizing the nondescript lands including that of the rooftops. A large part of the world population is living in peri-urban areas. The rich social and economic variability of peri-urban areas offers opportunities for the promotion of farming including edible green leaf production and thereby keeping harmony both between society and nature and between groups of people.

Strategies and Recommendations

1) Government should come up with proper policy including land bank creation and entitlement decision.
2) IPCC should incorporate ND lands as a strategic intervention for combating global warming and climate change.
3) University research programmers should be undertaken.
4) Agricultural research should reset its destination by incorporating NDL.

Chapter 2

FOOD CRISIS OF THE PLANET: A NON DESCRIPT PATHWAY TO FIGHT HUNGER AND MALNUTRITION*

ABSTRACT

Like nondescript plant biodiversity, the human diasporas can also be classified into descript and non-descript diasporas. Nondescript Diaspora critically means a section that is beyond the attention. Hunger, malnutrition and fragile immunity are the threats to the very existence of the vulnerable population. Chronic hunger is the silent killer that causes nutrient deficiency and poor immunity. The food crisis scenario is very alarming in south Asian sub continent, sub Saharan Africa and some of the Latin American countries. The problem beneath this food crisis is not merely low production or productivity. On the contrary, it is because of the scanty and uneven distribution of food among the population. One section of our population is starving and the other section is striving hard to lose weight. Poor immunity is also a problem that needs immediate look after and fruitful intervention. Many of the non descriptive plants habituated mainly in India are the rich sources of nutrients and are really effective in boosting up the immunity. Other than India, countries like

* Chapter Contributor: Anannya Chakraborty, PhD (Department of Agricultural Extension, Bidhan Chandra Krishi Viswavidyalaya, Mohanpur-741252, Nadia, West Bengal, India) to this chapter.

China and Germany are also taking necessary actions in extracting the medicinal and nutritional properties of these particular plants.

Keywords: chronic hunger, food crisis, hunger, malnutrition, immunity, non-descript plants

Introduction

Like nondescript plant biodiversity, the human diaspora can also be classified into descript and non-descript diaspora. Nondescript diaspora are those who are left beyond our attention, left beyond the welfare activities undertaken by the government, set aside beyond the purview of development initiatives. So we can classify them as the section of human population who are not known or taken care of. From the recent phenomenon we can find that the migrant workers are nondescript to their own village as well as to their erstwhile workplace or job place. No government projects, schemes, initiatives etc. are focused on the migrant laborers. Like the nondescript crops they are unattended and uncared of. While we are focusing on the nondescript leafy vegetable from the biological point of view, the same way we need to give an eye on the nondescript population within the social ecosystem. The other components of this nondescript population are the marginal and landless laborers, the farm women force, the daily wage earners, the domestic helps, the brick kiln workers and those who are already been marginalized in economic and social terms. So, this is the high time to take care of these nondescript population who are not in focus, (some ethnic groups may also be included). From this nondescript diaspora we can learn about the innovations of survival, the pristine knowledge system, the pragmetarian and survival instinct, the competition, conflict and the winner's instinct as well.

Theoretical Perspective

At one side there is a much proclaimed declaration on the technology driven growth and prosperity of the present civilization but the other side of this prosperity is bleak and disastrous which has got no match with the past centuries even. Out of around 7 billion population of the world, 1.5 billion are hungry; they don't have adequate access to food, if it is there, the quality doesn't stand anywhere near to fulfill their calorie requirement. In India, around 350 million people are living below the poverty line, and of them, 200 million people have fallen victim to moderate to extreme hunger indexes. 43 per cent of the new born babies are underweight while 60 per cent of the children are suffering from moderate to high level of anemia experiencing stunted growth (IFPRI report, 2017).

Global Scenario of Hunger and Food Prices

Driving forces behind soaring food prices are many and complex, with both supply-side and demand-side factors playing a part. Long-term structural trends and underlying growth in demand for food have coincided with short-term cyclical or temporary factors affecting food supply (Acharya, 2007).For the world community, hunger and food in security has been one of the highest priority issues ever since the World Food Summit (WFS) of 1996 had set a target of reducing the number of hungry and food insecure to half (842 million in the base period of 1990-92) by 2015. Later, the UN Millennium Declaration of 2000 set eight Millennium Development Goals (MDGs), the first of which was to reduce the proportion of hungry (from a base level of around 20 per cent) to half by 2015. Other seven MDGs were also closely linked to the first goal of reducing hunger (reducing malnutrition, IMR and MMR; increasing school enrolment; improving gender equality; increasing access to safe drinking water and sanitation; and combating HIV/AIDS, malaria and other diseases) (FAO,2005).Obviously, the MDG was milder than WFS target.

FAO's assessment for the period 2003-05 shows that the number of people suffering from chronic hunger worldwide has increased to 848 million, representing an increase of six million over the base period estimates. However, the proportion of hungry people (MDG) during 2003-05 came down to around 16 per cent. During this period, in Sub-Saharan Africa (SSA), the number of hungry people went up from 169 million to 212 million, but the proportion dropped from 34 per cent to 30 per cent. The rising food prices since 2006 have further aggravated the situation of hunger worldwide. The number of hungry in 2007 went up to 923 million (75 million plus over 2003-05 figure) and is likely to have increased further currently as the food price rise was much steeper in the early-2008. The proportion of hungry people is reported to have also increased to17 per cent in 2007. Thus, meeting the internationally agreed hunger-reduction goals in the few years remaining to 2015 is becoming an enormous challenge. At the regional level, the largest increases in the number of hungry people, as a result of rising food prices have taken place in Asia and Sub-Saharan Africa (SSA).The two regions combined already accounted for 750 million, or 89 percent, of the hungry people in the world in 2003-05. The rising food prices have driven an additional 41 million in Asia and 24 million in SSA below the hunger threshold. In other regions also, although the numbers are smaller, the hunger has increased due to rising food prices. Even in Latin America, which has shown a decade of steady progress towards the WFS target, there has been a sharp reversal of trend since 2006 (FAO, 2008).

In some international quarters, one of the major factors pointed out for rise in global food prices was increase in demand by the middle class in India (and also China), leading to food shortages. India's dietary pattern at least cannot substantiate this argument. During 2004-06, the average per capita consumption of cereals in India was 175 kg per year as against 953 kg in USA, 288 kg in China, and 316 kg in the world as a whole. Simplistic calculation would show that if the world adopts India's dietary pattern, only 56 per cent of the existing global cereals output is enough to feed the world.

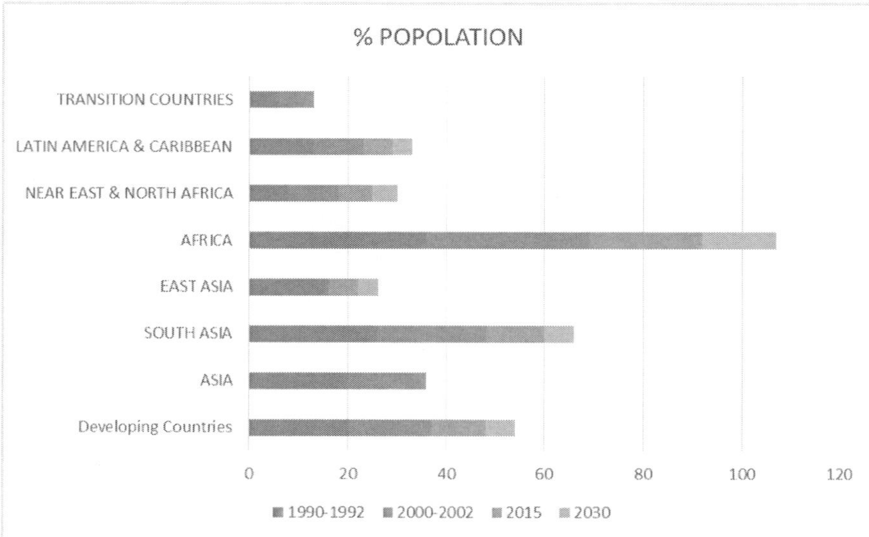

Figure 2.1. Decadal changes in percent population.

Food and Agricultural Organization is cautiously optimistic about the future. The progress in nutrition is expected to continue globally, in resonance with a reduction in poverty as projected by the World Bank. Although progress is expected to be slower than in the past, the incidence of undernourishment is expected to fall from 17 percent of the population of developing countries to 11 percent in 2015 and just 6 percent in 2030. Sub-Saharan Africa is the exception, however. The numbers of poor rose steeply in Sub-Saharan Africa during the 1990s and seem likely to continue to rise, at least over the near future.

Global Food Security: Approaches and Standing Schemes

Food security, at each the national and house levels, has been the focus of agricultural development strategy. The new strategy is based on food security on 3 key components, viz. (a) provision of associate degree improved technology package to the farmers; (b) delivery of contemporary farm inputs, technical power and institutional credit to the farmers; and (c) assurance of a remunerative selling and valuation setting for farmers. For achieving these objectives, many policy instruments were used that influenced production potential, and selling system of agricultural commodities in general and food merchandise in explicit. Some of the main policy instruments that mirror the policy package adopted by India are as follows:

i. Creation, strengthening and enlargement of the agricultural analysis system for developing and perfecting new production technologies for food grains and different agricultural commodities.
ii. Establishment, strengthening and enlargement of agricultural education and coaching system for agricultural extension employees and farmers for transfer of latest technologies at the farm level.
iii. Associate degree arrangement for the production and/or import and distribution of high-yielding farm inputs like improved seeds, fertilizers, agro- chemicals and improved implements/farm machines to the farmers.
iv. Coming up with and execution of major, medium, and small irrigation schemes for increasing the realm underneath irrigation.
v. Nationalization of business banks, creation of cooperative credit establishments, and reorientation of financial policy to increase institutional credit flow to the farmers.
vi. Creation and enlargement of physical and institutional infrastructure (primary market yards, roads, storage facilities, farmers' cooperatives and public sector organizations) for improvement of the selling system to handle and distribute the rising marketed surplus.
vii. Regulation of traders' exploitive selling practices through a series of legal and restrictive measures like licensing, levies, stocking limits and movement restrictions.

viii. Fixations of minimum support costs (MSPs) and arrangements for subsidy purchases and procure by public or cooperative agencies.
ix. Building-up and maintenance of buffer stocks of cereals and distribution of cereals through public distribution system (PDS).
x. Provision of food and input subsidies, express or implicit, for accommodative the conflicting objectives of cereal producers and shoppers in terms of costs.
xi. Canalization, licensing, minimum export costs (MEPs), and custom duties for maintaining value stability within fiscal market.

Dimensions of Food Security

Food security is achieved once 'all individuals in the least times have physical and economic access to food that's comfortable to fulfill dietary wants for a healthy and productive life'. During this sense, accomplishment of food security implies manufacturing (or importing) comfortable food and creating it accessible to all people throughout the year and on a property basis from year to year. Further, fulfilling dietary wants for a productive and healthy life implies physical and economic access of all individuals to wholesome food, consistent with every individual's demand. Food security so connotes freedom from hunger and deficiency disease.

The food security of a rustic must be assessed at 3 levels. First is that the availability of food at national level on property basis that depends on level and growth of food production, or adequate capability to import food (if availableness elsewhere is assured). Second is that the physical and economic access of all households to food. Physical access needs economical selling, transport, and storage system to hold the food at intervals a straightforward reach or at an inexpensive distance from human settlements (villages). Economic access of each house to food depends on its buying power and costs of food at that it is out there. And third is the employment of obtainable food by people that depends on intra-family allocation of food, and maintenance of affordable level of health of all people to consume and absorb needed level of food. Social factors like education, primary attention, gender bias, and role of ladies in house selections have an effect on food security at the individual level.

It is during this context that India tackled the hunger and food-insecurity scenario through each long and short measure. As a half of long strategy, it adopted a development strategy encompassing maintenance of adequate growth of national food production, and employment and incomes of plenty, improvement in selling infrastructure and access to education and first attention. These apart, the short strategy concerned selective market intervention and targeted distribution of backed food to cut back hunger and food insecurity. Further, because the organic process standing is additionally influenced by non-food factors like clean water and sanitation, it absolutely was recognized that spotlight to those aspects can facilitate translate food security into smart nutrition.

Household and Individual Food-Security

i. The challenge of measuring food security became increasingly thorny during the 1990s, when the number of conceptual elements proliferated in widely accepted definitions. There have been commendable recent advances in the development of simple, valid measures of "food access" (Coates, 2013). There is a substantial improvement in physical access of households to food in numerous components of the world that was contributed by many factors. First, the share of rice, that is a lot of geographically distributed, in total staple food, continues to be quite high at around 45 per cent. Second, the growth of network of public distribution system helped in reaching cereals to deficit and geographically tough regions (hilly and desert areas). And third, growth of road networks, creation of primary market yards, and building-up of storage facilities within the rural areas inflated physical access of rural households to food even in otherwise deficit areas.

ii. Improved convenience of staple food at the declining real costs has contributed to improved nutritional-security. There has been a shift by the farmers from the low-yielding coarse cereals to non-cereal food product since the center of eighties, a truth that has lay is the green revolution helped to extend production and convenience of sugar, fruits, vegetables, spices, milk, eggs, meat and fish/fish product. Because the production growth of all these food things was significantly beyond the population growth, per capita production of wholesome foods went up considerably. Asian nation currently produces 58 Mt fruits, 112 Mt vegetables, 102 Mt milk, 46 billion eggs and 7 Mt fish.

iii. Despite growth within the convenience of cereals and decline in their relative costs vis-à-vis incomes, the per capita consumption of cereals has conjointly cared-for come by recent years. The decline in consumption has been chiseler in coarse cereals, and has occurred even among the lowest 30 per cent of shoppers, reflective a shift towards a lot of wholesome foods like fruits, vegetables and placental mammal product. Semi permanent knowledge from National Sample Survey Organization (NSSO) has conjointly indicated a declining trend within the per capita consumption of cereals in each rural and urban area, in the midst of a decrease within the proportion of expenditure on cereals and a rise therein on milk, meat, eggs, fruits and vegetables. This shows improvement within the organic process levels.

iv. A paper of Coates et al. (2006) hypothesizes that there is a common "core" to the household food insecurity experience that goes beyond insufficient food quantity and that transcends culture. The paper for the first time employs an exploratory approach to identify cross-cultural commonalities of the food insecurity experience as captured in 22 scales and related ethnographies derived from 15 different countries. Economic poorness is Associate in nursing vital issue poignant food security at the house level. Over the years, the incidence of poorness, in each rural and concrete area, has declined significantly. The proportion of population below poverty level declined from 51 per cent in 1977-78 to 39 per cent in 1987-88, to 26 per cent in 1999-00 and to 22 per cent in 2004-05.

FOOD POLICY AND SUPPORT FOR FARMERS

Many a government round the globe actively intervenes and controls the operation of their agricultural markets. The ways that they intervene and also the reasons they are doing therefore rely in massive half on the wealth of the country. Governments in poor accumulation countries habitually impose worth controls to stay food costs unnaturally low. They are doing therefore to realize favor with their lot of politically powerful urban residents. Although various peasant farmers do not organize politically, and thus, have abundant less political power than their urban brethren. The irony of this example is that by unnaturally depressing the worth of food, accumulation governments cut back incentives for farmers to supply and cut back the supply of food from autochthonic sources.

In continent the prevalence of the lower incentives of farmers is that the most as a result of this can be the one continent that expertise systematically declining per capita food production within the postcolonial amount. In several African nations, state promoting board's area unit granted a legal monopoly to shop for agricultural product from farmers and to sell them to domestic shoppers and in export markets. Such boards typically pay farmers solely one third to half the domestic shopper worth or the export worth. The shrinking range of farmers makes organizing in interest teams easier. What is more, political redistricting typically lags behind the shift in population to the cities. As a result, rural districts typically have a lot of legislative representatives and luxuriate in larger political power than their numbers would recommend. Farmers use this power to hunt higher and a lot of stable farm costs via legislation or act. But the great political organization is not the sole reason that farmers reach obtaining governments to lift their costs. A second reason is that farmer's area unit typically viewed as underprivileged. Rural communities lack several of the amenities that cities have and since labor productivity is mostly lower in agriculture than in producing, wage rates area unit lower. Also, technological amendment tends to expand agricultural production quicker than consumption, reducing the worth of farm product. In 1870, as an example, the worth of wheat was over eleven greenbacks per bushel. Today, it's solely regarding four greenbacks per bushel, a drop of over sixty per cent though shoppers gain by paying lower costs, the incomes of farmers drop. As labor leaves agriculture in search of upper financial gain within the cities, the reduced provider of farmers' causes the remaining farmers' incomes to rise back to their previous level. This will take years, though. Another reason governments intervene to support farm costs is that they typically area unit volatile. Climatic conditions, over those farmers don't have any management, area unit a vital determinant of what proportion a farmer harvests in a very given year. The ensuing variability of production within the face of comparatively stable demand causes farm costs, and farmers' incomes, to vary from year to year. This could cause economic hardship for farm families in a very unhealthy year. It should conjointly cause farmers to travel bankrupt as a result of fashionable farming needs massive investments in specialized facilities and instrumentation.

Forms of Subsidy

1) If the farmers of a rustic don't turn out enough quantity of grain to satisfy the country's consumption functions, then it's inevitable for the govt. to try and do the worth support. The remainder is formed up through imports. In these cases the country merely imposes associate degree tariff or quota till the domestic value rises to the required level. Growers receive the upper value, and customers pay the upper value for each import and for domestic production. For instance, within the mid-eighties, once the globe market value of

sugar was four cents per pound, USA import quotas were therefore limiting that the domestic wholesale value exceeded twenty cents per pound.

2) A rustic manufacturing additional crops than its demand, the worth supporting is hard for its government and needs a considerable support of beaurocrats. Government a minimum legal value below that an honest can be sold-out seldom works. Therefore rather than government minimum costs, governments typically try and raise costs unnaturally by limiting production. Every farmer could also be issued a quota that stipulates what quantity he will sell in an exceedingly given year. This can be done through with peanuts within USA and milk in North American nation. Limiting production effectively cartelizes the trade, and also the government enforces the pool.
3) When there is an increase of the worth by the government, the beneficiaries' area units of the farmers are enhanced. As a result of their insufficiency, the quotas straightaway withstand worth. All future entrants should purchase a quota to achieve the correct to sell the merchandise. That raises the investment needed to become a farmer and also the price of production. Once the first quotas area units sold-out to new farmers, those farmers become a robust lobbying force against ever leaving behind quotas.
4) This "set-aside" approach seldom is extremely effective at supporting agricultural costs. Farmers don't seem to be stupid; they put aside their least productive land first. what is more, a policy that makes artificial insufficiency of land induces farmers to accentuate their production practices on every acre that continues to be in production, raising its yield. Therefore unless terribly giant reductions in are needed, set-asides alone seldom cut back production greatly. Moreover, aggravating production typically needs heavier doses of fertilizer and agricultural chemicals, with probably adverse environmental consequences. For instance, farmers within the European Economic Community (EC), wherever support costs for grain area unit more than those within

the us, use over doubly the maximum amount fertilizer per acre than U.S. farmers. As a result, variety of northern European countries area unit encountering elevated levels of fertilizer nutrients in their groundwater.

5) The foremost usual U.S. approach to support of associate degree marketable agricultural product is to make an agency to shop for any amount of a product offered by the countries farmers at the secure "support price."That keeps market costs at or close to the support value. Inside the US Department of Agriculture, this agency is termed the Goods Credit Corporation (GCC). Support costs should be in the course of import quotas. Otherwise, foreign producers would sell their product within the U.S. market as long because the U.S. value exceeded the worth they may get elsewhere. If that happened, the U.S. government would finally end up guaranteeing the U.S. value to farmers round the world. A U.S. example is dairy farm product that the cardinal buys to support the farm value of milk. Quotas limit imports of dairy farm product to buy 3% of consumption. The cardinal disposes of the commodities it buys in ways in which won't displace market demand and depress the domestic market value. For instance, dairy farm product area unit typically given away to low-income folks, within the college lunch program, and as economic aid. A variant of this policy is meant to stabilize market costs. The cardinal buys grain at the support value, stores it, and releases it back to the market if the market value rises to a prescribed trigger level of, say, 140% of the support value. During this manner the policy protects growers against the danger of low costs however additionally protects customers against remarkably high costs. This sort of state program will offer some protection against wide swings in costs if the acquisition at about 75% of a five-year moving average of market costs (leaving the very best variety and also the lowest variety out of the calculation). The price between acquisition and unharnessed value ought to cowl the price of in operation the buffer stock program.

6) But the farm organizations typically do lobbing to boost the acquisition and unharnessed costs, in order that "stabilization policy" becomes subsidy policy. Once this happens, government inventories tend to rise while not limit till the stabilization agency exhausts its take into account shopping for the merchandise. This can be precisely what happened once the Federal Farm Board, the precursor of the cardinal, tried to support the U.S. costs of wheat and cotton in 1929. At that time the agency must subsidize the export of the inventories, with the taxpayers learning the loss on the operation.
7) USA has presently taken a combined approach for subsidy together with the supply of loans. At harvest the cardinal offers grain farmers nine-month loans capable their production times the support value. The support value is termed the "loan rate." The cardinal accepts the grain as collateral for the loan. If, throughout the term of the loan, the rises higher than the support price, farmers repay the loans with interest and sell the grain within the market. If the market value remains at or below the loan rate, farmers forfeit the grain to the cardinal, keep the money, and don't have any more obligations. Such loans area unit referred to as nonrecourse loans, which means that the loaner has no claim on the receiver on the far side the collateral (in this case the crop).

Some Programs for Eradicating Hunger and Deficiency Disease: The Worldwide Strategy

The Zero Hunger Challenge

It was first planned at the international organization Conference on property Development in Brazil in 2012. Secretary-General Ban Ki-moon's Zero Hunger Challenge was aiming for a future wherever each individual has adequate nutrition. This needs comprehensive efforts to make sure that each man, ladies and kid get pleasure from their Right to Adequate Food; ladies area unit empowered; priority is given to family farming; and food systems area unit property to be resilient.

Zero Loss or Waste of Food

Eliminating hunger involves investments in agriculture, rural development, good work, social protection and equality of chance. It will create a significant contribution to peace and stability and to the reduction of economic condition. It will contribute to raised nutrition for all – particularly ladies from the start of gestation and kids underneath the age of 2.

Four Steps towards Achieving Zero Hunger per Word Economic Forum

Put Those Furthest Behind First

To understand the total potential of our globalized economy, national governments should expand social protection schemes for the foremost vulnerable. Providing this chance for just economic process can raise the buying power of the poorest a pair of billion folks that successively can produce progressive demand, generate new jobs and jump-start native economies. Finance in inclusive development isn't simply the correct factor to try and do, it makes smart business sense too.

Pave the Road from Farm to Market

Access to cheap, wholesome food for everybody – all seven billion folks – is important. We have a tendency to should pioneer and invest in creating our provide chains additional economical by developing property sturdy markets. To support these markets, we have a tendency to should additionally improve rural infrastructure, significantly roads, storage and electrification, guaranteeing farmers' ability to succeed in a wider client base.

Reduce Food Waste

Of the four billion metric tons of food we have a tendency to turn out annually, one third is wasted, cost accounting the worldwide economy nearly $750 billion annually. In developed countries, food is commonly wasted on the plate, whereas in developing countries it's lost throughout production, as crops go unused or unprocessed as a result of poor storage or as a result of the farmers cannot get their merchandise to promote.

Encourage a Property Form of Crops

These days, across the world, four crops (rice, wheat, corn and soy) represent hour of all calories consumed. To handle the challenges of temperature change, food convenience and access, we'd like to assist farmers explore and determine an additional various vary of crops. Agricultural crop diversity can probably offer communities with the nutrients needed for healthy growth and a lively *modus vivendi*. To realize this objective we have a tendency to educate farmers within the cultivation of those crops, which incorporates guaranteed access to the requisite tools and skills. Even as necessary, we have a tendency to build a client marketplace for these foods by educating communities regarding the nutritional importance of uptaking a good food.

Supplementary Feeding Program (SFP)

According to Austin (1981) and Austin and Zeitlin (1981) analysis, the "take-home" system requires less frequent attendance, fewer dispensaries and health centers and involves fewer personnel in the distribution. Although the "on-the-spot" feeding programs have shown a high percentage of substitution, they seem to ensure the intake of food by the beneficiaries (Beaton & Ghassemi, 1982). Because the SFP thinks about, the most objectives of the supplementary feeding programs were to boost the organic process standing and health of the beneficiaries, to stimulate regular group action at health centers, and conjointly to introduce nutrition education. In sight of the actual fact that health is joined to instructional, environmental and demographic variables, it's necessary to integrate supplementary feeding programs with different activities. Some fascinating intervention studies are undertaken in Central American nation by INCAP. One in all them investigated the influence of nutrition supplements on kid development, and therefore the impact of medical services provided by health auxiliaries on health conditions, over associate eight year amount (1979-1987). Another study analyzing the interaction between nutrition and infection was created in Asian nation (the "Narangwal Project") providing nutrition supplements and education in one space, health care in another and each nutrition and health services in one more. Another programs conjointly focused on the utilization of various food-stuffs, as within the Philippines (nutripack); Democratic Socialist Republic of Sri Lanka (Thripoha), and in Asian country (IPRPD products).

In Asian nation an identical program is running named Supplementary Nutrition Program (SNF). The target of SNP is to alleviate or stop deficiency disease among vulnerable kids below half dozen years and expectant or nursing mothers. The program was launched in 1975 and currently covers 4.8 million mothers and 22.9 million kids through a network of 4200 comes covering seventy five per cent of development blocks within the rural areas and 273 slum pockets within the urban areas. According to organic process wants, hot meals or snacks, together with different things (vitamins and iron tablets, etc.) square measure provided through service (Anganwadi) centers established within the neighborhood with a neighborhood girl because the in-charge.

World Food Program(WFP)

WFP was established in 1963 once the 1960 Food and Agriculture Organization (FAO) Conference, once St. George McGovern, director of the US Food for Peace Program, projected establishing a three-sided food aid program. The WFP was formally established in 1963 by the Food and Agriculture Organization of the United Nations and therefore the global organization General Assembly on a three-year experimental basis. In 1965, the program was extended to an unbroken basis.

The WFP strives to eradicate hunger and deficiency disease, with the final word goal in mind of eliminating the necessity for food aid itself. Its objectives square measure to:

a) "Save lives and shield livelihoods in emergencies"
b) "Support food security and nutrition and (re)build livelihoods in fragile settings and following emergencies"
c) "Reduce risk and modify individuals, communities and countries to satisfy their own food and nutrition needs"
d) "Reduce under-nutrition and break the inter-generational cycle of hunger"
e) "Zero Hunger in 2030."

The activities of WFP square measure widespread and quite vital. Within the year 2013, the WFP reached 80.9 million individuals in 75 countries and provided 3.1 million tons of food, together with nutritionally enriched ready-to-use therapeutic foods. 7.8 million Unnourished kids received special organic process support in 2013, and 18.6 million kids received faculty meals or net rations. Once more in 2015, the WFP reached 76.7 million individuals in 81 countries. In emergencies, quite 50 million individuals were reached so as to boost their nutrition and food security. Faculty meals were provided to 17.4 million kids, serving to keep them at school to make sure uninterrupted access to education. The WFP has scaled up its use of money and vouchers as food help tools. Money or voucher program supported 7.9 million individuals in 2013. Within the same year, the WFP purchased food in 91 countries; 86 of that food came from developing countries. In 2017 WFP launched the Building Blocks Program. It aims to distribute money-for-food help to Syrian refugees in Jordan. The project uses block chain technology to digitalize identities and permit refugees to receive food with eye scanning. The WFP conjointly unloads humanitarian aid at the Freeport of Monrovia national capital port throughout Joint Task Force Liberia. Among its different activities, the

WFP has coordinated the five-year Purchase for Progress (P4P) pilot program. Launched in Sept 2008, P4P assists land farmers by giving them opportunities to access agricultural markets and to become competitive players within the marketplace. The project spanned across 20 countries in Africa, Asia, and geographic area and trained 800,000 farmers in improved agricultural production, post-harvest handling, quality assurance, cluster selling, agricultural finance, and acquiring with the WFP. The project resulted in 366,000 metric loads of food made and generated quite US$148 million in financial gain for its farmers. The WFP focuses its food help on those that square measure most susceptible to hunger, that most often means that ladies, children, the sick and therefore the older. In fact, a part of the response to the 2010 Haiti earthquake consisted of distributing food aid solely to ladies as expertise engineered up over virtually 5 decades of operating in emergency things has incontestable that giving food solely to ladies helps to make sure that it's unfold equally among all social unit members. School-feeding and/or net ration program in seventy one countries facilitate students concentrate on their studies and encourage oldsters to send their kids, particularly women, to school (WFP, 2015).

HUNGER AND POLITICS

"When I die, don't engineer a monument on American state. Don't bestow American state degrees from nice universities. Simply dress the naked. Say that I attempted to deal with the homeless. Let individuals say that I attempted to feed the hungry"…Martin theologize King.

World War II gave birth to a replacement international politico-economic order that was later represented as embedded liberalism. It heavily funded the United Nation's development programs, and later the efforts of different three-sided organizations just like the International fund (IMF) and therefore the UN agency (WB).The recently established global organization became a number one player in coordinating the world fight against hunger. The world organization has 3 agencies that job to push food security and agricultural development: the Food and Agriculture

Organization (FAO), the planet Food Programs (WFP) and therefore the International Fund for Agricultural Development (IFAD). Food and Agriculture Organization of the United Nations is that the world's agricultural information agency, providing policy and technical help to developing countries to push food security, nutrition and property agricultural production, notably in rural areas. WFP’s key mission is to deliver food into the hands of the hungry poor. The agency steps in throughout emergencies and uses food to help recovery once emergencies. Its long run approaches to hunger helps the transition from recovery to development. IFAD, with its information of rural poorness and exclusive concentrate on poor rural individuals, styles and implements programs to assist those individuals access the assets, services and opportunities they have to beat poorness. Following victorious post World War II reconstruction of Germany and Japan, the IMF and WB began to show their attention to the developing world. An excellent several civil society actors were conjointly active in making an attempt to combat hunger, particularly once the late Nineteen Seventies once world media began to bring the plight of starving individuals in places like Federal Democratic Republic of Ethiopia to wider attention. Most vital of all, particularly within the late 60s and early 70s, the revolution helped improved agricultural technology propagate throughout the planet. Thus USA began to alter its approach to the matter of world hunger from concerning the middle 50s. Important members of the administration became less obsessed on strategies they saw as promoting associate over reliance on the state, as they feared that may assist the unfold of communism. Despite this read, throughout the 60s post war era hunger inside the USA was overshadowed by hunger in Europe and Asia. John F. Kennedy did indeed use government Order to double the quantity of commodities on the market from the excess artifact program additionally as initiated the pilot legal tender Program that later became permanent in 1964. By the 80s, the previous agreement in favor of moderate government intervention had been displaced across the western world. The IMF and UN agency specially began to push market-based solutions. In cases wherever countries became captivated with the IMF, they generally forced national governments to prioritize debt repayments and sharply cut public services. This generally had a negative result on efforts to combat hunger (FAO, IFAD & WFP, 2014).

The issue of food sovereignty was brought forward by Organizations like Food initial, it additional claimed that each country on earth (with the potential minor exceptions of some city-states) has spare agricultural capability to feed its own individuals, however that the "free trade" economic order, that from the late 1970s to concerning 2008 had been related to such establishments because the UN agency and World Bank, had prevented this from happening.

The World Bank claimed to be a vicinity of the answer to hunger, declarative that the simplest approach for countries to interrupt the cycle of financial condition and hunger was to make export-led economies that offer the money suggests that to shop for foodstuffs on the globe market. However, within the early twenty first century the globe Bank and UN agency became less dogmatic concerning promoting free market reforms. They more and more came to the read that government intervention will have a job to play, which it may be advisable for governments to support food security with policies favorable to domestic agriculture, even for countries that don't have a Comparative advantage in this space. As of 2012, the globe Bank remains active in serving to governments to intervene against hunger.

Until a minimum of the 1980s—and, to associate extent, the 1990s—the downside of demand Olympian offer was a dominant educational outlook with reference to world hunger. Projected solutions usually centered on boosting food production, and generally on contraception. There have been exceptions to the current, while early because the Forties, Lord Boyd-Orr, the primary head of the UN's United Nations agency, had perceived hunger as for the most part a drag of distribution, and drew up comprehensive plans to correct this. Few in agreement with him at the time, however, and he resigned when failing to secure support for his plans from the US and nice UK. In 1998, Amartya Sen won an award partially for demonstrating that hunger in times isn't usually the merchandise of an absence of food. Rather, hunger typically arises from food distribution issues, or from governmental policies within the developed and developing world. It's since been generally accepted that world hunger results from problems with the distribution further because the production of food. In 2007 and 2008, quickly increasing food costs caused a worldwide food crisis, increasing the numbers plagued by hunger by over 100 million. Food riots erupted in many dozen countries; in a minimum of 2 cases, Haiti and Madagascar, this crystal rectifier to the falling of governments. A second world food crisis unrolled because of the spike in food costs currently 2010 and early 2011. Fewer food riots occurred, due partially to larger availableness of food stock piles for relief.

Non-Descript Approaches to Overcome Crisis of Hunger

A framework for non-descript pathway has been imminent to get rid of the claustrophobia of hunger-poverty and voice.

The Stress-1 and the Panacea

The land for agriculture is squeezing first. By 2050, 60 per cent of the global population is going to be confined in city ecology. The roof top spaces are needed to be included into non-descript land masses. The urban parks may have to undergo a transformation from ornamental biology to nutrition garden with ornamental properties. The colorful capsicums, for example, can serve both the purposes. Same way, we have to go for roadside plantation of shrubs, creepers, vines and trees having both beauty and nutritive utility. E.g., planting of drumstick, amla, amaranths, sunflower and so on can be prescribed as roadside beautification. Ornamentals should be redefined with utility and health care.

The Stress-2 and the Panacea

A prolongs and skewed food habit over centuries followed by millions of people for a geographical expanse is certain to unleash stress for ecosystem services. For example, a prolonged and traditional dependence on rice, wheat, potato, sugarcane can be lethal for an agro-ecosystem. The genes will be fatigued, resources will be extorted and the efficiency will be shattered. A non-descript food habit including roots, leaves, vines, flowers, weeds, pods, mollusk, icthyiofaunal scores in overlooked water regimes, non-timber forest product can be mainstreamed as food for human being. Else, the ecological resilience will be jeopardized.

The Stress-3 and the Panacea

Declaring agriculture as industry, branding agriculture as the sub-sector is nothing worse than a racial profiling. A segregationist's approach to agriculture has damaged the better tomorrow of agriculture. There can't be any denial that farmers' need bonus, health care support, pension policies and, above all, an industrial employee status. As hospitals need 24 hours dedicated power supply through hot lines, same way, the farm and farm families need a dedicated power grid to ensure food production efficiency round the year. Is it a utopia, and if so, every non-descript concept has been reborn into neo-utopias.

The Stress-4 and the Panacea

Over centuries, our farmers are treated as shuttle cocks, and constant have been hit and punched by market and income uncertainties. The grand design of these uncertainties are traceable when production is up to ensure food security, the market price goes down to invite grand uncertainty, often gets reflected through a distress sell. So, they must be supported by Equal Monthly Income (EMI) against a pledged production and pledged income by and for the famers. You may call it a non-descript approach to save farming diasporas from an early extinction.

The Stress-5 and the Panacea

Folk gene library, run by wise farmers, farm women, school teachers, scientists, observers and student activists, will go a long way in saving, conserving and regenerating non-descript genotypes in every village. School premises are the best locations where there will be a perfect blending of formal and folk knowledge, as if a plant or vegetables library will keep on opening up the wonders of wild vines, fodders, plants, fungi, honey bees, or local fish and livestock breeds. A simple, traditional, overlooked pond in a village can offer a wonder of fish biodiversity, just like the Beethoven's hilarious symphony. This is a beautiful non-descript approach to get resources blended with rural intelligence and creativity for an ecological function functional over a protractile period.

Throughout history, parts of the world's population have usually old sustained periods of hunger. In several cases, this food offer disruptions were caused by war, plagues, and adverse weather or generally even artificial. For the primary few decades when warfare II, technological progress and increased political cooperation has created the struggle against hunger and famine quiet easier. Whereas progress was uneven, by 2000 the threat of maximum hunger subsided for several of the world's individuals. Per the WFP (World Food Policy) statistics show that, "Some 795 million individuals within the world don't have enough food to steer a healthy active life. That is concerning one in 9 individuals on earth. The overwhelming majority of the world's hungry individuals sleep in developing countries, where 12.9% of the population is malnourished."

Until 2006, the common international worth of food had been for the most part stable for many decades. Within the closing months of 2006, however, costs began to rise at a pace. By 2008, rice had tripled in worth in some regions, and this severely affected the developing countries. Food costs fell in early 2009, however rose to a different record margin in 2011, and has since diminished slightly. The 2008 worldwide money crisis additional increased the amount of individuals plagued by hunger, together with dramatic will increase even in advanced economies like nice UK, the Euro zone and therefore the USA.

The Millennium Development Goals enclosed a commitment to an additional five hundredth reduction within the proportion of the world's population who are suffering from extreme hunger by 2015. Although as of 2012, this target appeared tough to realize, because of inflation in food costs. However, in late 2012 the UN's Food and Agriculture Organization (FAO) declared it's still potential to hit the target with spare effort. In 2013, the FAO estimated that 842 million individuals are malnourished (i.e.12% of the worldwide population). Deficiency disease may be an explanation for death for quite three.1 million kids below five per annum. UNICEF estimates three hundred million kids head to bed hungry every night; which 8000 kids below the age of five are calculable to die because of deficiency disease a day.

Malnutrition and Poor Immunity: The Utility of Non Descript Plants

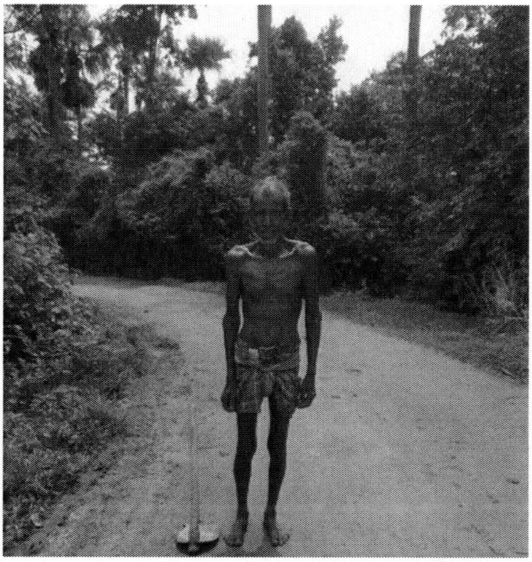

This is a picture of a farmer of Bankura District, West Bengal (India) focusing on the real and ground level scenario. These farmers alongwith the farmers in general, who are our bread givers, are basically living on the edge of hunger, malnutrition, poor immunity and poverty. According to FAO, nearly 22% of the Indian farmers are suffering from malnutrition. India's malnutrition is of global proportion. UNICEF estimated that India was home to 42% of the developing world's children who are underweight (low weight for age) and 32% of those who were stunted (low height for age). Now as far as the global malnutrition scenario is concerned 462 million people are underweight, 47 million children under 5 years of age are wasted, 14.3 million are severely wasted and 144 million are stunted (UNICEF, 2009).

Now as far as the immunity status is concerned, as per Indian context, rural people have a stronger immune system than that of the urban or semi urban population. This difference is basically because of the life style, food habits, stress, pollution predominance etc. As the urbanization and modernization goes, the prevalence of pathogenesis is also goes in a robust speed. A study conducted by Chakraborty and Acharya (2019) in some of the chosen districts of West Bengal, India revealed that the rural community is lacking of having so called immunity boosters or fancy medications, but they are blessed with a vast collection of nature's treasures in the form of many nondescript plants like swamp weed/kulekhara (*Hygrophilous spinosa*), Malabar nut/basak (*Justicia adhatoda*), stinkvine/gandal (*Paederia foetida*) and many other plants with medicinal values. These herbs are really boosting up their immunity and giving them inner strength that they are less vulnerable to the diseases. These non descriptive plants are not only the habitat of India, but also these have a larger family in many other south East Asian countries like Bangladesh, Sri Lanka, Thailand etc.

In many regions of the world, people still rely on the medicinal plants for their primary healthcare and boosting up their immunity. This is mainly popular in rural communities in Africa, some parts of Asia, central and South America where the traditional use of the plants and their medicinal property is quite common and accessible. In the year 2016 the Chinese government announced that their aim is to incorporate Traditional Chinese Medicine (TCM) in to their healthcare system by the year 2020. In Germany it was found in a study that 90% of the population use herbal medicines. USA is also very keen to use these herbal medicines in a modernized and improvised form. They have spent 17 billion US dollars in 2000 on traditional herbal medicines. The world Health Organization has estimated an annual global market for herbal medicines to be worth 60 billion US dollars.

INFERENCES

So from this overall discussion we can clearly point out that-

1) The problem of hunger malnutrition and poor immunity is not because of the lower production of food; on the contrary, it is basically because of the uneven distribution of food among the overall population.
2) There is a huge social, economic and technical gap in between the privileged and under privileged population of the world and the only way to lower the gap is mass education.
3) Hunger and malnutrition is creating a nexus with the silence (poor communication) of the people who are living at the receiving end.
4) Poor immunity is a problem for both privileged and under privileged population of the world.
5) The traditional medicinal herbs and the non descriptive plants are really an awesome source of immunity boost up and nutrition as well. The super powers like USA and China are now have started investing on the redemption of the traditional and non descriptive plants of higher medicinal values.

Chapter 3

BIODIVERSITY AND INDIGENOUS KNOWLEDGE OF NON-DESCRIPT GREEN LEAVES IN ATTAINING CLEAN AND HEALTHY FOOD*

ABSTRACT

Indigenous knowledge and biodiversity are two complementary parts to help in human ecology development. This chapter will portray the relation role of indigenous knowledge and the rural communities in India and other parts of the world in nurturing knowledge based on a sophisticated understanding of their environment, conservation and sustenance of surrounding natural resources and bio-diversity, local minor plant species, passing and exchanging information through generations. However, this indigenous knowledge is often ignored and maligned by outsiders. Though, today, however, some of the governments and international development agencies are recognizing that local-level knowledge and organizations provide field for participatory approaches. Domesticated and non-domesticated plants are getting importance in Indian sub-continent as well as in most other parts of the world, providing an important opportunity for systematic in situ

* Chapter Contributor: Riti Chatterjee, PhD Scholar (SRF), (Department of Agricultural Extension, Bidhan Chandra Krishi Viswavidyalaya, Mohanpur-741252, Nadia, West Bengal, India) to this chapter.

conservation of genetic resources. Besides, there are some possible extension approaches that will make the local people aware of the valuable indigenous leafy vegetables, though, much of this knowledge is at as much risk of being lost as is the case with biodiversity. Hence, both wild and cultivated leafy vegetable plants must be nurtured in a sustainable manner to stop these from becoming extinct. Here, geographical indication tag and intellectual property right or biodiversity act are some of the strongest supporters in documentation of the wild, minor plant species. Using the present study as a baseline, if the nutrient compositions and other nutra-medicinal properties of the locally available leafy vegetables, particularly under-utilized species could be traced out, it might be possible to alleviate malnutrition in the world.

Keywords: biodiversity, indigenous knowledge, minor vegetables, rural people, wild plants

INTRODUCTION

Tropical ecosystems have undergone huge changes in terms of habitat loss, fragmentation and shifting towards commercial agriculture. Hence, traditional agro-forestry systems (on-farm conservation) could be a promising option for ensuring environmental resilience. Such systems can help eradicate ecosystem degradation while giving access to food and economic opportunities to rural people. However, traditional knowledge is important in the course of a broad range of questions related to the relationship between humans and nature. This traditional knowledge is not any static entity for any set of skills in any culture, it is seen that people often change their techniques when easier methods become available. This is because, different communities from different places of the world perceive and interact with nature differently and acquire different types of environmental knowledge. This is also same in case of some leafy vegetables, which theoretically can be termed as ‘non-descript’, as those do not have any commercial production; they are just used for home consumption without conscious nurture. These types of leafy vegetables are the most prominent species in homestead gardens beyond the eyesight.

In some West African countries, local farmers keep some key species near their houses. The literature on the importance, diversity and management of minor and non-descript leafy vegetables is however relatively poor. Maximum of the available studies concentrate on the biodiversity and its potential to conserve the threatened species and wild crop species. This chapter will try to elucidate traditional knowledge on minor leafy vegetables and how this knowledge shapes the choice of species and could sustainable conservation of biodiversity.

BIO-DIVERSITY IN ATTAINING HEALTHY FOOD

Biodiversity, the term sometimes sound complicated, but it's a simple concept, the existence of many different types of plants and animals makes the world a healthier and more productive place. A mix of genetics, species and habitats allows earth's ecosystems to cope up with challenges like population growth and climate change. Biodiversity is important to us because it plays a critical role in food and nutrition security, and at the same time for human health. Over the years, reduced access to land and natural resources, environmental degradation, climate change, globalization, and urbanization of agricultural production have caused a shift in diets and lifestyles. In many communities, the decreasing availability of diverse local foods and the increase of industrialized foods have caused a shift away from traditional food resources toward commercial and convenience foods. This trend has been associated with negative health impacts like obesity, diabetes, heart disease, high blood pressure and other diet-related disorders.

Understanding and maintaining the key relationships between peoples and their cultures, biodiversity, traditional livelihood and knowledge systems are crucial factors in ensuring sustainable food and nutrition security. Though many traditional food practices have already been lost, however, still there is a chance to recover indigenous food systems, so that we can recapitulate long-standing knowledge and practices to conserve biodiversity and obtain benefit from it in the present and in the future time. Because restoring knowledge and awareness is one the first step of utilizing and revitalizing the biodiversity.

It is seen that in Burkina Faso, rural women carefully collect the fruit, leaves and roots of native plants such as the baobab tree, red sorrel leaves, kapok leaves and tiger nut tubers for use in their families' diets. These plants can be supplements to their diet, where agricultural grains like rice, wheat, millet, and sorghum and together complement the nutritional spectrum. And more than 800 species of edible wild plants have been indexed so far. It has found from Burkina Faso that local knowledge of food biodiversity sometimes needs aspects of gender dynamics. Men and women may be responsible for different crops, or varieties of crops, or be responsible for different tasks related to one crop.

As women often have detailed knowledge and preferences of traditional and non-descript crops, as well as play a central role in seed selection, seed saving, and use of wild plants for food. Also sometimes they handle charge of minor food crops used for home consumption, rituals and medicinal properties. Men, on the other hand, tend to be in charge of the cultivation of commercial crops. The expansion of men's commercial enterprises may result in a reduction of women's crops, leading to a decrease in the availability of local plants for food and medicine. This shift signifies a possible declination in nutritional status, reduction in local plant diversity and overall environmental resilience. Therefore, it is necessary to take both angles of agriculture into consideration when striving for a sustainable future.

One of the most important questions of the preset world is how to increase food production to meet the sustainable food security while improving biodiversity and reducing the strain on the environment. Good governance, enabling frameworks, and stewardship incentives are needed to facilitate mainstreaming of biodiversity. Hence, we have to commit to improving and maintaining biodiversity today (FAO 2018).

TRADITIONAL AGRICULTURE AND CROP DIVERSITY

Practices governing the use of biological and natural resources may be observed in the context of forest product use, traditional water management, landholding patterns, agricultural practices, fisheries, use of commons, etc. Communities possess different forms of traditional operative frameworks of rights, powers, concessions and obligations relating to the use and management of the biological and natural resources. In the present times, indigenous communities have different types of dependence on the biological and natural resources in their vicinity. Over the years, in the Indian context, the direct dependence of several communities and families has decreased over these resources as they have found new options and means of livelihoods in their own communities and sometimes result migration to other places. Yet there are numerous communities in all parts of the country even today that are directly and largely dependent for their sustenance and survival on these natural resources. These communities have a stake in conserving and using the resources in a sustainable manner (Gadgil 2003).

Indigenous People

Before knowing indigenous technologies, traditions, one must be acquainted with the indigenous peoples. According to the World Bank (1991), indigenous peoples are "social groups with a social and cultural identity distinct from the dominant society that makes vulnerability to being disadvantaged by the development process." Where, the UN has no universally accepted definition but thinks "indigenous communities, peoples and nations are those which, having a historical continuity with pre-invasion and pre-colonial societies that developed within their territories, think themselves different from other sections of the society now prevailing in those territories" (Cobo 1987). Indigenous peoples are people living in an area within a nation or state, prior to the formation of a nation/state, but they may identify with it; and have maintained a great part of their solely distinct linguistic, cultural, social and organizational characteristics. These differentiate them in some degree from the surrounding populations and dominant culture of the concerned nation/state.

Indigenous/Traditional/Folk Knowledge

Indigenous knowledge, which is also referred to as 'traditional knowledge', 'local knowledge', 'folk knowledge' etc., refers to that body of knowledge developed by local and indigenous communities over time in response to the needs of their specific local environment (Gadgil 2003). These indigenous knowledge generally pass on to other generations and have enable indigenous people to survive, manage their natural resources and the ecosystems surrounding them like animals, plants, rivers, seas, natural environment, economic, cultural and political organization. Knowledge of these aforesaid elements then forms a set of interacting units known as 'indigenous coping systems'. Indigenous knowledge system refers to the set of interactions between the economic, ecological, political, and social, environments within a group or groups with a strong identity, drawing existence from local resources through patterned behaviours that are transmitted from generation to generations to cope, to sustain with change. These patterns are also sustained in micro level with differentiated responsibilities that ensure the group's continuous survival. Unfortunately, these systems are eroding very fast due to colonialism, commercialization, globalization and modernization, lack of efficient indexing, breakdown of traditional family structure and function, and so on. This indigenous knowledge is mainly of a practical nature, particularly in such fields as agriculture, fisheries, health, horticulture, forestry and environmental management in general.

Social Factors and Indigenous Knowledge

Indigenous people always find appropriate ways to guarantee the survival of the minor and neglected species as they play key role in their daily needs. Though, there may be some differences that could be linked to specificity of the native vegetation, while there may be some propagation of some species through human migrations.

Photo 1. Different locally available non-descript leafy vegetables.

Geographically closed sociolinguistic groups share similar species. In the Guineo-Congolean zone, Fon and Holli shared highest similarity value, where Anii, Nago and Ditamari, Gourmantché had more species respectively in the Sudano-Guinean and the Sudanian zones. Thus, culture, cultural link and environment influenced selection and resource use in this way. Hence, prominent sociolinguistic groups of each phytochorological zone share secular linkages. Indeed, wedding arrangements are customary between Fon and Holli people. This could have encouraged knowledge sharing among these sociolinguistic groups. Furthermore, they found that, as, most surveyed people were illiterate; they could have remained much close to their culture. As such, knowledge could have been transferred through generations easily and with little erosion. There were no differences between usage types of men and women within the same phytochorological zone. However, noticeable differences were observed among phytochorological zones with respect to type of use and species raised in the homestead lands.

Indigenous Knowledge on Some Locally Available Non-Descript Vegetables

A large part of the world population, fulfill their nutritional requirements by consuming various types of leafy vegetables (Singh and Arora 1978). Leafy vegetables are generally referred to leaves of any plants used as vegetables, sometimes accompanied by petioles and shoots (Mazumder and Sarkar 2019).

Few people know about this rich culinary treasure of local and seasonal edible greens that grow across India. From treating common ailments to adding variation to daily diet, these indigenous super-foods have been consumed in rural India for centuries. Leafy vegetables can also be used to earn some small income, as the local people can also sell them in their local markets. Spinach (*palak*), fenugreek (*methi*), amaranth (*note*) and mustard greens (*sarson*) are some of the commonly used and known leafy greens —they are generally called as *saag*. Above all, it is a universally acknowledged that leafy green vegetables come under nature's best nutrition supplements. They come with iron, calcium, vitamins, antioxidants and fibre, and are vital components of a healthy and balanced diet.

Thus, there is a real need of maintaining proper records of leafy vegetables and their status of conservation through agricultural practices. Some of the important leafy vegetables and surrounding indigenous knowledge has been presented below:

Table 3.1. Detailed information on different locally available non-descript leafy vegetables

Name of the non-descript leafy vegetables				Use	Range of use	Cultivation status	Major economical parts	Indigenous knowledge (if any)
Common name	English name	Botanical name	Family					
Puinnya Saag	Swampweed	*Hygrophila polysperma* (Roxb.) T. Anders.	Acanthaceae	Cooked as vegetable	Limited	Wild	Leaf	Found in Indian sub-continent as well as in USA
Luchipata saag	Peperomia	*Peperomia pellucida*	Piperaceae	Eaten raw as medicinal plant, also cooked as vegetables	Moderate	Wild	Leaf	Effective against cough and cold, prostate problems and many inflammations.
Bichu Buti/ sisunaak saag	Stinging Nettle	*Urtica dioica*	Urticaceae	Cooked asvegetable	Limited	Wild (throughout the Himalayan region of India)	Leaf	According to folklore, a broth made of nettle leaves was the only food consumed by the renowned Tibetan mystic Milarepa during his weeks-long retreat in the forest.

Table 3.1. (Continued)

Name of the non-descript leafy vegetables				**Use**	**Range of use**	**Cultivation status**	**Major economical parts**	**Indigenous knowledge (if any)**
Common name	English name	Botanical name	Family					
Kalmegh	Green Chiretta	*Andrographis paniculata* (Burm.f.) Nees	Acanthaceae	Eaten raw as medicinal plant against stomach problem	Wide	Wild and locally cultivated	Leaf	Effective against worms, if taken at early morning in empty stomach
Basak	Malabar nut	*Justicia adhatoda* L.	Acanthaceae	Leaf extract used as oral medicine for common cold and cough	Wide	Locally cultivated	Leaf	
Kulekhara	Hygrophila	*Hygrophila auriculata* (Schumach.) Heine	Acanthaceae	Eaten raw as medicinal plant	Wide	Wild andcultivated	Leaf	Most effective for adolescent girls as it boosts haemoglobin
Lalsaag	Red Amaranthus	*Amaranthus tricolor* L.	Amaranthaceae	Cooked asvegetable	Wide	Commercially cultivated	Leaf	
Sadanote	WhiteAmaranthus	*Amaranthus blitum* subsp.*oleraceus* (L.) Costea	Amaranthaceae	Cooked asvegetable	Wide	Commercially cultivated	Leaf	

Name of the non-descript leafy vegetables				**Use**	**Range of use**	**Cultivation status**	**Major economical parts**	**Indigenous knowledge (if any)**
Common name	English name	Botanical name	Family					
Lingru	fiddlehead fern	*Matteuccia struthiopteris*	Onocleaceae	Cooked as vegetables	Limited	Locally cultivated	Leaf	Widely consumed saagin Uttarakhand and Himachal Pradesh, where it grows abundantly in the wild. The locals of Uttarakhand, cook it into a curd-based curry, while in the Kullu valley it is mostly consumed as a pickle.
Palong saag	Spinach	*Spinacia oleracea* L.	Amaranthaceae	Cooked as vegetable	Wide	Commercially cultivated	Leaf	
Note saag	Amaranthas	*Alternanthera sessilis* (L.) R.Br. ex DC.	Amaranthaceae	Cooked as vegetable	Limited	Wild	Leaf	
Thankuni saag	Centella	*Centella asiatica* (L.) Urb.	Apiaceae	Eaten raw as medicinal plant	Wide	Wild	Leaf	Very effective in cough and cold
Amrit saag/ Tengechi saag	Indian sorrel	*Oxalis corniculata*	Oxalidaceae	Eaten raw as medicinal plant	Wide	Wild	Leaf	Used as folk medicine

Table 3.1. (Continued)

Name of the non-descript leafy vegetables				**Use**	**Range of use**	**Cultivation status**	**Major economical parts**	**Indigenous knowledge (if any)**
Common name	English name	Botanical name	Family					
Kunth	Cyathula	*Cyathula* sp.	Amaranthaceae	Cooked as boiled or vegetable	Wide	Wild	Leaf	Used as folk medicine
Kochu	Colocasia	*Colocasia esculenta* (L.) Schoti	Araceae	Cooked as vegetable	Wide	Wild and commercially cultivated	Leaf and Rhizome	
Mankochu	Alocasia	*Alocasia macrorrhizos* (L.) G.Don	Araceae	Cooked as vegetable	Wide	Wild and commercially cultivated	Leaf and Rhizome	
Oal	Amorphophallusbul bifer	*Amorphophallus bulbifer* (Roxb.) Blume	Araceae	Cooked as vegetable	Moderate	Wild and Commercially cultivated	Leaf and Rhizome	
Bonpalong	Sonchus	*Sonchus arvensis* L.	Asteraceae	Cooked as vegetable	Moderate	Wild	Leaf	Mainly popular for sour dishes
Dheki	Diplaziumescu-lentum	*Diplazium esculentum* (Retz.) Sw.	Athyriaceae	Cooked as vegetable	Wide	Wild and Commercially cultivated	Leaf	
Puisaag	Basella	*Basella alba* L.	Basellaceae	Cooked as vegetable	Wide	Wild and commercially cultivated	Leaf	

Name of the non-descript leafy vegetables				**Use**	**Range of use**	**Cultivation status**	**Major economical parts**	**Indigenous knowledge (if any)**
Common name	English name	Botanical name	Family					
Bothuasaag	Chenopodium	*Chenopodium album* L.	Chenopodiaceae	Cooked or boiled as vegetable	Wide	Wild	Leaf	
Dhudla Saag	Hewitiamalabarica	*Hewitia malabarica* (L.) Suresh	Convolvulaceae	Cooked or boiled as vegetable	Limited	Wild	Leaf	
Haldi Patta	Turmeric	*Curcuma longa*	Zingiberaceae	Cooked	Limited	Wild and commercially cultivated	Leaf and rhizome	In the coastal regions of India, turmeric leaves are traditionally used to make herbal concoctions to ward off cold, fever and other respiratory allergies. During the monsoon months in Goa, turmeric leaves are used to make a rice dish called Patoli.

Table 3.1. (Continued)

Name of the non-descript leafy vegetables				**Use**	**Range of use**	**Cultivation status**	**Major economical parts**	**Indigenous knowledge (if any)**
Common name	English name	Botanical name	Family					
Telakucha	Ivy gourd	*Coccinea indica*	Cucurbitaceae	Cooked as vegetable	MODERATE	Wild	Leaf And fruit	Used traditionally in treatment of intestinal worms, skin diseases, etc.
Kolmi	Water spinach	*Ipomoea aquatica* Forssk.	Convolvulaceae	Cooked as vegetable	WIDE	Wild	Leaf	
Ramtulsi	Clove basil	*Ocimum gratissimum* L.	Lamiaceae	Leaves are Consumed as raw	WIDE	Wild and locally cultivated	Leaf	Used as home remedy in the treatment of cough and cold.
Bantulsi	Jungle tulsi	*Ocimum basilicum* L.	Lamiaceae	Leaves are used as home remedy in the treatment of cough and cold	WIDE	Wild and locally culivated	Leaf	Stems and branches are also used as fuel, then the ash is collected and stored for 5-6 days, then put in warm water and use as detergent.

Name of the non-descript leafy vegetables				**Use**	**Range of use**	**Cultivation status**	**Major economical parts**	**Indigenous knowledge (if any)**
Common name	English name	Botanical name	Family					
Krisna Tulsi	Holy basil	*Ocimum tenuiflorum* L.	Lamiaceae	Leaves are Consumed as raw	Wide	Wild andlocallycultivated	Leaf	Used as home remedy in the treatment of cough and cold.
Sushni	Four leaf clover/european waterclover	*Marsilea quadrifolia*	Marsileaceae	Cooked as vegetable	Wide	Wild	Leaf	This plant is said to be anti-inflammatory, diuretic, depurative, febrifuge and refrigerant. Tribal people also used to treat snakebite and applied to abscesses.
Dhol Pata/ Kanshira	Benghal dayflower/tropical spiderwort/ wandering Jew/Commelina	*Commelina benghalensis*	Commelinaceae	Cooked as vegetable	Wide	Wild	Leaf	Its use as afamine food in India has been recorded. In southeast Asia and Africa it is used as fodder and also medicinally as apoultice.

Table 3.1. (Continued)

Name of the non-descript leafy vegetables				**Use**	**Range of use**	**Cultivation status**	**Major economical parts**	**Indigenous knowledge (if any)**
Common name	English name	Botanical name	Family					
Baro Nunia saag	Portulaca			Leaves are used as flavouring substance	Limited	Wild	Leaf	
Gima saag/dima saag	Chickweed	*Glinus oppositifolius* (L.) Aug.DC.	Molluginaceae	Cooked asvegetable	Wide	Wild	Leaf	Effective against liver ailments, fever, cold socks, skin diseases.
Brahmi saag	Bacopa monnieri	*Bacopa monnieri* (L.) Wetist.	Plantaginaceae	Cooked as vegetable	Wide	Locally cultivated	Leaf	Believe to improve brain functions, strengthening memory, make anxiety free and promotes sound sleep.
Gandal	Paederia	*Paederia foetida*	Rubiaceae	Used as raw extract	Limited	Wild	Root and leaf	Traditional medicine used in wound and bone healing.
Uchchhepata	Bittergourd leaf	*Momordica charantia*	Cucurbitaceae	Fried	Wide	Wild and commercially cultivated	Leaf and fruits	Good against itching

Source: Mazumder and Sarkar 2019 and authors own knowledge.

Some Possible Extension Approaches in Conservation of Indigenous Knowledge

There must be some driving agents that will foster the preservation of indigenous knowledge and the on-farm diversity of the non-descript and minor leafy vegetables through the generations:

- Focus on local knowledge, resource, innovation and assets.
- Empower the farmers and community institutions as legitimate actors in maintaining the national agricultural biodiversity system.
- Social learning platform to cope with vulnerability in present climate change context.
- Promote good governance for ensuring ecological sustainability through conservation of local community knowledge and natural resources.

However, informal agricultural research and development systems parallel those of national government must provide opportunity for national agricultural research and extension services to work with the creative interests and activities of farmers and other rural people who are taking care of indigenous plants generation after generation. Global network of regional and national indigenous knowledge resource centers are taking initiatives in documenting the historical and contemporary indigenous knowledge of numerous ethnic groups around the world, though, the knowledge is always at risk of being lost and same happens in case of biodiversity (Linden 1991). There are number of steps in preserving and documenting indigenous knowledge with the help of extension approaches:

1) enhancing community awareness
2) understanding local biodiversity, social networks and institutions
3) capacity building of community institutions
4) establishing institutional working modalities
5) consolidating community roles in planning and implementation

6) community monitoring and evaluation, and
7) social learning and scaling up for community collaboration efforts.

GEOGRAPHICAL INDICATION (GI)

Many countries hold the view that the existing legal framework, particularly the intellectual property system could be used for improved protection to indigenous knowledge and that under certain circumstances; geographical indications (GIs) could be a particularly important way of protecting traditional products. The ability to extend the life of GIs indefinitely and the possibility of collective ownership of such rights suggest that they may be especially suitable for protecting indigenous knowledge. Geographical indications (GIs) may induce protection of the collective rights of the rural and indigenous communities in their indigenous knowledge, ensuring that the entire community that has passed it on with required refinement over generations, stand to benefit from the knowledge and is not locked up as the private property of one individual group or community.

Other advantages of GIs are that knowledge remains in the public domain, the scope of protection is limited to controlling the class and/ or location of people who may use the protected indication and the rights as long as the product-place link is maintained. Holders of a GI do not have the right to assign the indication, thus, preventing its transfer to non-local producers. The latter provision is important for protection of indigenous knowledge and to ensure that it does not pass on to the hands of those who are not holders of the knowledge.

In case of leafy vegetables, the first is the basic level or a basic standard of protection where all geographical indications must be protected against use which would mislead the public or constitute an act of unfair competition. This obligation is met with in most countries (Commission on Intellectual Property Rights 2004).

THE BIOLOGICAL DIVERSITY ACT, 2002

The Biological Diversity Act, 2002 was enacted with the objectives of conservation of biological diversity, sustainable use of its components and fair and equitable sharing of benefits arising from the utilization of genetic resources. The Act and the Rules made there under in 2004 constitute part of the Indian attempt to operationalize the two most important stipulations of the Convention on Biological Diversity. The first is the sovereign right of countries of origin over their genetic and biological resources. The other is the acceptance of the need to share benefits flowing from commercial utilization of biological resources with holders of indigenous knowledge. As yet, there is no proper resolution at the international level as to how these will be implemented in view of the fact that the normal Intellectual Property Rights and TRIPS provisions do not stipulate any sharing of benefits with holders of knowledge nor is the sovereign right of countries of origin over their genetic and biological resources acknowledged (Commission on Intellectual Property Rights 2004).

CONCLUSION

The non-descript leafy vegetable plants and their utilization is well recognized by the local communities. It is said that the majority of the local inhabitants were dependent on wild vegetation for under-utilized leafy vegetables but over-utilized leafy vegetable plants were commercially cultivated in many places. However, in some cases over-utilization of such wild or non-descript leafy vegetables may affect the ecological diversity and create threats to the vegetation. Therefore, both wild and cultivated leafy vegetable plants need to be used in a sustainable manner. Taking the present study as a baseline, if the nutrient compositions and other nutra-medicinal properties of the leafy vegetables, particularly under-utilized species could be determined, it would be possible to alleviate poverty and malnutrition in different corners of the world.

Chapter 4

IMMUNITY AND NUTRITION*

ABSTRACT

Food is the medicine. Food can provide medicine to cure and immunity to fight. The role of intake of iron, zinc, phosphorous, sodium, calcium, vitamins and natural proteins help generating antibodies against antigens. This has become so essential especially during pandemic disasters. People at large should be aware of immune systems and its science of building up in our body system. The limitation of calorie rich traditional food is that it ignores with blasphemy the need for immunological build up process. The strength of non descript crops is that it enters our body system, keeps moving with higher compliances with our metabolic functions and triggers up immune build-up process. A kind of auto immunization process makes up capable of fighting against contamination and infectivity.

Keywords: immunity, antibody, medicine, protein, vitamins

INTRODUCTION

Antigen produces various microscopic organisms and toxic products. These antigens are responsible for various diseases. Microscopic

---

* Chapter Contributor: Rama Das, (Assistant Professor, Department of Food & Nutrition, Barrackpore Rastraguru Surendranath College, West Bengal, India) to this chapter.

organisms are called microbes or germs. The germs infect our body through inhaling air, through water or through foodstuff etc. Novel Corona Virus case was first reported in Wuhan, Hubei Province, China in December, 2019 and spread across the world (Thevarajan et al. 2020). Human body has a system to combat the infection by germs in various ways. Our body also fights against antigen to produce antibody. This protective system of our body is called Immune System. Immune system is made up of different types of cells, viz., lymphocyte – T-lymphocyte and B-lymphocyte; null cell; mononuclear phagocytes – monocyte and macrophage; granular leucocyte – neutrophil, eosinophil, basophil; mast cell; dendritic cell, etc. These cells protect our body from various diseases by different ways.

PROTEIN

Antibodies are protein in nature. It is made up of two pairs of polypeptide chains and each pair contains four hundred amino acids. So, amino acid or protein is the main constituent of antibody and we should include protein in our daily diet to produce and for proper functioning of antibodies.

Table 4.1. Recommended dietary allowances of protein

Source	**Protein (gm/100gms)**
Fish	22-24
Meat	16.5-23
Egg	13-14
Milk	2-3.7
Soya chunk	52
Wheat flour	10.57
Bengal gram whole	21.55
Ground nut	23.65
Green gram dal	23.88
Barley	10.94

IFCT 2017.

Nutrients acting directly on cells of the immune system may change the capacity of immune cells to recognize a foreign stimuli. The concentrated interaction of amino acids, vitamins and minerals all of which must originated in the diet is required for the synthesis of those proteins, i.e., immunoglobulins, enzymes and lymphokines which participated in immune responses. Protein is the most important nutrient for boosting our immune system. Daily requirement of protein of an adult is 1 gm per Kg of body weight while for children it is 1.5 gm/Kg body weight (Table 4.1).

RETINOL

Vitamin A may be considered to have the capacity to influence several features of the immune system. It is needed for maintaining the normal health and function of epithelial layers which delivers the first way to defend invading microorganisms. Vitamin A or its metabolites are required for regulation of humoral immune response (antibodies) and cellular immune response which involves the direct killing of infected cell. Even it is also required to revise the course of immune response in cases of acute respiratory infection, pneumonia and bronchiolitis and thus having impact on the outcome of respiratory infection. We can meet our vitamin A requirement from the foodstuffs given in Table 4.2.

Table 4.2. Foodstuffs with vitamin A content

FOOD STUFF	mcg RAE PER SERVINGS
Beef liver,pan fried,3 ounces	6582
Sweet potato,baked in skin, 1whole	1403
Spinach,frozen,boiled,½ cup	573
Carrots,raw,½ cup	459
Cheese, ricotta, part skim,1cup	263
Herring, Atlantic, pickled, 3 ounces	219
Mangoes,raw, 1 whole	112
Egg,hard-boiled,1 large	75
Apricots, dried, sulfured, 10 halves	63
Broccoli,boiled,½ cup	60

IFCT 2017.

VITAMIN D

It is found from several researches that vitamin D has an immense role on calcium and bone homeostasis and the Vitamin D receptor is expressed on B cells, T cells and antigen presenting cells (immune cells), hence, they bear the capacity to synthesize the active vitamin D metabolite. It has the capability to modulate the innate and adaptive immune response. Apart from having impact on calcium and bone homeostasis, Vitamin D bears more vital role in different aspect. B Cells, T Cells and antigen presenting cells are immune cells on which Vitamin D receptor is expressed. These immunologic cells can be able to synthesize the active vitamin D metabolite.

Vitamin D having capacity to work in an autocrine mode in a local immunologic environment is able to control the innate and adaptive immune responses. Deficiency in vitamin D is linked to enhance the probability of autoimmunity and that of proneness of getting infection. Again in case of autoimmune diseases where the ameliorative effects of vitamin D acquired responses from immune cells, vitamin D deficient individuals with autoimmune disease may be benefitted with supplementation of vitamin D apart from the benefits of vitamin D on bone and calcium homeostasis.

The protective immunity of our body has been maintained by the immune system which forms, apart from maintaining tolerance to self, the defensive protection from invading foreign organisms. Recent studies have unearthed the impact of deficit of vitamin D on the immune system and such deficiency may otherwise promote the increased vulnerability to infection and often to a diathesis particularly in a genetically inclined host to autoimmunity. Multiple cross-sectional studies have been undergone with respect to presence of increased infection with lower levels of vitamin Deity has been established through comparative studies over 19000 subject for 1988-1994 that individuals (other than those having sufficient level of vitamin D) have noticed self-report of recent upper respiratory tract infection when measured as having lower levels of vitamin D (<30mg/ml) in spite of corrected variables of season, age, gender, body mass and race.

Even data on cross-sectional study on level of serum vitamin D held over 800 military recruits in Finland have shown that significantly loss of more days from active duty for secondary to upper respiratory infections were noticed in comparison to those recruited having higher level of vitamin D.

Though it is fact that rates of seasonal infections are not constant since lowest in the summer and highest in the winter and that vitamin D levels also vary over the year, lower level of serum vitamin D is found to have linkage with predominance of infection in each season (Cynthia Aranow 2011).

We can meet our vitamin D requirement from the food given in Table:

Table 4.3. Foodstuffs with vitamin D content

Food stuff	**Vitamin D IU/100g**
Cod liver oil	10000
Shrimp	150
Liver(lamb)	120
Mackerel	120
Cod	85
Butter	35
Egg yolk	25
Beef stick	13
Cheese	12
Corn oil	9

IFCT 2017.

VITAMIN C

Vitamin C having pleiotropic functions linked with its capability to donate electrons may be considered as an essential micronutrient for humans. It plays the role of a potent antioxidant and that of a cofactor for a group of biosynthetic and gene regulatory enzymes. By means of giving various cellular functions of the innate as well as adaptive immune system, it ultimately contributes to the immunity. Vitamin C frames the potential protectiveness to counter environmental oxidative stress by way of supporting epithelial barrier function to resist pathogens stimulating/encouraging the oxidant scavenging activity of the skin. Being accumulated in phagocytic cells (e.g., neutrophils), Vitamin C promotes chemotaxis, phagocytosis, generation of reactive oxygen species and thereby stands effective for killing microorganisms. Even Vitamin C is responsible for decreasing necrosis and potential damage of the tissues since it plays role in apoptosis and in wiping out the spent neutrophils from sites of infection by macrophages (Carr and Maggini 2017).

We can meet our vitamin C requirement from the food given below:

Table 4.4. Foodstuffs with vitamin C content

Food stuff	Vitamin D IU/100g
Amla	600
Gooseberry	252
Guava,White flesh	214
Whole Wheat Flakes	207
Cashew Fruit	180
Parsley	133
Capsicum, Green	123
Agathi Leaves	121
Drumstick Leaves	108
Green Chillies	94.07

IFCT 2017.

ZINC

Zinc is considered as an antioxidant. It is essential for maintenance of protein structure and for the metabolism of protein and nucleic acids. It also plays a vital role for synthesis of immunoglobulin. The daily requirement of zinc is 9-12 mg per day. Zinc from animal products is more readily absorbed than plant products. Dry Oyster contains 1 mg/gram. Meat, poultry, eggs and dairy products are the best sources. The main sources of zinc of pulses and vegetables. These contains 224 mg/100 g. Nuts also contain high amount of zinc.

The function of zinc with respect to the body immune system may be compared to the gatekeeper. Adequate functioning of innate vis-à-vis adaptive immunity is also being controlled by zinc flux, zinc wave and homeostatic zinc signals. Severe functional impairment of immune system has been noticed due to deficit of zinc while an excess in zinc also promotes impairment of the immune system. So, it is obvious that a balanced zinc homeostasis is essentially required to stand a defense towards invading pathogens as well as to shield our body against an over reactive immune system resulting into auto-immune diseases, chronic inflammation or allergies. Hence, the function of all immune cells is virtually dependent to zinc (Zalewski 1996).

We can meet our zinc requirement from the food given below:

Table 4.5. Foodstuffs with zinc content

Food stuff	mg PER SERVINGS
Oysters, cooked, breaded and fried,3 ounces	74
Crab, Alaska king, cooked,3 ounces	6.5
Lobster,cooked,3 ounces	2.9
Pumpkin seeds,dried,1ounce	2.2
Cashews,dry,roasted,1ounce	1.6
Almonds,dry,roasted,1ounce	0.9
Peas,green,frozen,cooked,½ cup	0.5
Flounder or sole,cooked,3 ounces	0.3
Yogurt,fruit,low fat,8 ounces	1.7
Chickpeas,cooked, ½ cups	1.3

IFCT 2017.

Selenium

Further, being a vital constituent of selenocysteine-containing protein, Selenium is linked with most of the functions of cell biochemistry. Hence, to promote the immune system of the body, selenium may be considered as having enormous value. Instance may be cited from the fact that selenium, as a selenocystein, is a component of the active site of the antioxidant glutathione peroxidases that have active role in protecting neutrophils from oxygen-derived radicals yielded to destroy/exterminate ingested foreign organisms (Wolonciej et al. 2016).

The immune system is typically an intensely complex group of procedures functioning together to defense against the invading pathogens and malignancy. However, functions of various aspects of the immune system relates to Selenium.

We can meet our selenium requirement from the food given below:

Table 4.6. Foodstuffs with selenium content

Food stuff	mcg PER SERVING
Brazil nuts,1ounce(6-8nuts)	544
Tuna, yellow fin, cooked, dry heat,3 ounces	92
Halibut, cooked, dry heat, 3 ounces	45
Ham, roasted, 3 ounces	42
Shrimp, canned, 3 ounces	33
Cottage cheese, 1% milk fat,1 cup	20
Rice, brown, long-grain, cooked, 1 cup	19
Egg, hard-boiled, 1 large	15
Bread, whole wheat, 1slice	13
Lentils, boiled, 1cup	6

IFCT, 2017.

Plants being rich in flavonoids, vitamin C or the carotenoids, are capable of promoting immune function. Role as mild anti-inflammatory and as an immune-stimulant have been observed among the flavonoid-rich herbs. It has already been established that activity of lymphocytes, phagocytosis and interferon production have been boosted by those herbs. To cite a common instance among such remarkable plants having impact over immune system, name of garlic may be quoted that has the capability to deploy an immune-potentiating impact by promoting natural killer cell activity. It has been studied that potential of garlic as an immune modifier is surprisingly assertive and it is able to conserve the homeostatis of immune system as it contains a higher concentration of sulfur compounds considered as having therapeutic effects. Further, the chemicals present in garlic are established to be beneficial in treatment of cancer, diabetes, atherosclerosis and hyperlipidemia (Khodadadi 2015).

The sample menu chart fulfilled our daily needs to meet up nutritional requirement and the menu provides nutrients given below in Table 4.9.

Table 4.7. Recommended dietary allowances

	Source of goal[a]	**Child 1-3**	**Female 4-8**	**Male 4-8**	**Female 9-13**	**Male 9-13**	**Female 14-18**	**Male 14-18**
Calorie level(s) assessed		1000	1200	1400,1600	1600	1800	1800	2200,2800,3200
Macronutrients								
Protein, g	RDA	13	19	19	34	34	46	52
Protein, %kcal	AMDR	5-20	10-30	10-30	10-30	10-30	10-30	10-30
Carbohydrate, g	RDA	130	130	130	130	130	130	130
Carbohydrate, % kcal	AMDR	45-65	45-65	45-65	45-65	45-65	45-65	45-65
Dietary fibre, g	14g/1000kcal	14	16.8	19.6	22.4	25.2	25.2	30.8
Added sugars,%kcal	DGA	<10%	<10%	<10%	<10%	<10%	<10%	<10%
Total fat, % kcal	AMDR	30-40	25-35	25-35	25-35	25-35	25-35	25-35
Saturated fat, % kcal	DGA	<10%	<10%	<10%	<10%	<10%	<10%	<10%
Linolenic Acid, g	AI	7	10	10	10	12	11	16
Linolenic Acid, g	AI	0.7	0.9	0.9	1	1.2	1.1	1.6
Minerals								
Calcium , mg	RDA	700	1000	1000	1300	1300	1300	1300
Iron, mg	RDA	7	10	10	8	8	15	11
Magnesium, mg	RDA	80	130	130	240	240	360	410
Phosphorus, mg	RDA	460	500	500	1250	1250	1250	1250
Potassium, mg	AI	3000	3800	3800	4500	4500	4700	4700
Sodium, mg	UL	1500	1900	1900	2200	2200	2300	2300
Zinc, mg	RDA	3	5	5	8	8	9	11
Copper, mcg	RDA	340	440	440	700	700	890	890
Manganese, mg	AI	1.2	1.5	1.5	1.6	1.9	1.6	2.2
Selenium, mcg	RDA	20	30	30	40	40	55	55
Vitamins								

Table 4.7. (Continued)

	Source of goal[a]	**Child 1-3**	**Female 4-8**	**Male 4-8**	**Female 9-13**	**Male 9-13**	**Female 14-18**	**Male 14-18**
Vitamin A, mcg RAE	RDA	300	400	400	600	600	700	900
Vitamin E, mg AT	RDA	6	7	7	11	11	15	15
Vitamin D, IU	RDA	600	600	600	600	600	600	600
Vitamin C, mg	RDA	15	25	25	45	45	65	75
Thiamin, mg	RDA	0.5	0.6	0.6	0.9	0.9	1	1.2
Riboflavin, mg	RDA	0.5	0.6	0.6	0.9	0.9	1	1.3
Niacin, mg	RDA	6	8	8	12	12	14	16
Vitamin B_6, mg	RDA	0.5	0.6	0.6	1	1	1.2	1.3
Vitamin B_{12}, mcg	RDA	0.9	1.2	1.2	1.8	1.8	2.4	2.4
Choline, mg	AI	200	250	250	375	375	400	550
Vitamin K, mcg	AI	30	55	55	60	60	75	75
Folate, mcg DFE	RDA	150	200	200	300	300	400	400

[a] RDA=Recommended Dietary Allowance, AI =Adequate Intake, UL= Tolerable Upper Intake Level,

AMDR = Acceptable Macronutrient Distribution Range, DGA= *2015-2020Dietary Guidelines*recommended limit; 14 g fiber per 1000 kcal = basis for AI for fiber.

[b] Calcium RDA for males ages 71+ years is 1200 mg.

[c] Vitamin D RDA for males and females ages 71+ years is 800 IU.(The National Academies Press; 2006 and 2010)

Table 4.8. One day sample menu

Time	Item	Amount	
		Male	Female
Morning	Tea/Coffee	1 cup	1 cup
Breakfast	Bread Toast	3 slices	2 slices
	Sprouts	1 small bowl (20 gm)	1 small bowl (20 gm)
	Egg (Hen)	1 pc(50 gm)	1 pc(50 gm)
	Milk	100 ml	100 ml
Midmorning	Orange	1 pc(100 gm)	1 pc(100 gm)
	Almond/ Walnut	5 gm	5 gm
Lunch	Rice	150 gm	100 gm
	Pulses	20 gm	20 gm
	Leafy Vegetables	100 gm	100 gm
	Mixed Vegetables	200 gm	200 gm
	Fish Curry	1 pc (50 gm)	1 pc (50 gm)
	Curd	100 gm	100 gm
Afternoon/ Evening	Rice Flakes (Poha)	1 cup (100 gm)	1 cup (75 gm)
	Tea		
Dinner	Chapati	4 pcs	3 pcs
	Mixed Vegetables	200 gm	200 gm
	Paneer Curry	100 gm	100 gm

Notes:

- For vegetarian diet 1 portion of meat/fish/egg can be replaced by 1 portion of pulse
- Total Sugar: 20 gm
- Total Visible Fat: 20 gm
- This diet plan fulfils respective Recommended Dietary Allowances (RDA) for Adult Sedentary Workers

IFCT, 2017.

Table 4.9. Calculation of nutritional value for men

Energy Kcal	Carbohydrate(gm)	Protein (gm)	Fat(gm)	Fiber (gm)	Vitamin B_2(mg)	Vitamin B_6(mg)	Vitamin B_9(µg)	Vitamin C(mg)	Vitamin A(µg)	Carotenoids(µg)	VitaminD_2& D_3(µg)	Iron (mg)	Zinc (mg)	Calcium(mg)
2188.52	258.19	70.26	41.79	25	35.39	0.91	694.5	145.51	126.8	13458.49	17.59 & 0.37	14.6	6.93	862.2

IFCT, 2017.

Table 4.10. Calculation of nutritional value for women

Energy Kcal	Carbohydrate(gm)	Protein(gm)	Fat(gm)	Fibre(gm)	Vitamin B_2(mg)	Vitamin B_6(mg)	Vitamin B_9(µg)	Vitamin C(mg)	Vitamin A(µg)	Carotinoids(µg)	Vitamin D_2& D_3(µg)	Iron (mg)	Zinc (mg)	Calcium(mg)
1910.98	213.16	63.96	41.79	31.82	0.88	49.56	684.27	132.15	126.8	13458.49	17.59 & 0.37	14.6	6.93	862.2

IFCT, 2017.

Chapter 5

COMMON LEAFY GREENS

1. Common name: Amrul, Creeping wood sorrel, Sleeping beauty
Scientific name: *Oxalis corniculata*
Family: Oxalidaceae

INTRODUCTION

Oxalis corniculata is an annual or short-lived perennial plant. It grows only upto a height of about 5 – 10 cm, but the roots spread to form a mat of growth with about 30 cm or more width. The plant is generally harvested from the wild for food, medicine and various other commodities. It is an excellent therapeutic agent regarded for thousands of years and we continue to rely on them for health care needs. Plants packed with a wide range of ingredients can be effectively used in traditional medicine to treat chronic as well as infectious diseases.

Habitat

Anthropogenic meadows and fields are the natural habitat. It spreads rapidly by seed and underground bulbils and can quickly become a nuisance. It usually remains as common weed of barren lands but can be cultivated as pot plants in nurseries.

Medicinal Properties

The whole plant is anthelmintic, antiphlogistic (anti-inflammatory), astringent, depurative, diuretic, emmenagogue, febrifuge, lithontripic, stomachic and styptic. It is used in the treatment of influenza, fever, urinary tract infections, enteritis, diarrhoea, traumatic injuries, sprains and poisonous snake bites. The plant is a good source of vitamin C and is used as an antiscorbutic in the treatment of scurvy. It can also cure cough, cold, arthritis, dysentery, etc. The juice of the plant, mixed with butter, is applied to muscular swellings, boils and pimples. An infusion can be used as a wash to rid children of hookworms. The leaves are used as an antidote to poisoning by the seeds of Datura spp, arsenic and mercury and the leaf juice is used against insect bites, burns and skin eruptions. It has an antibacterial activity. An infusion of leaves is used to remove opacities of the cornea and is dropped into the eyes for itching lids. A decoction of leaves is used as a gargle.

It has significant wound healing activity by producing an increase in wound contraction rate, wound breaking and significant decreases in epithelization period. It inhibits the tumor growth in ascitic and solid tumor models. It has moderately significant antibacterial activity when compared with standard streptomycin and antifungal potency against *Aspergillus niger* and 10.7% against *Phaseosphaerella theae* oxalates and other unknown inhibitory substances may also be responsible in some kinds of allelopathic activity.

OTHER USES

Leaves are used as raw or cooked. They are added to salads and cooked as a potherb with other milder flavoured greens. The leaves are available all year round unless the winter is very cold, they have a pleasantly sour taste, but are very small and fiddly to harvest. The leaves contain about 86% water, 2.3% protein, 0.8% fat, 8.2% carbohydrate, 150 mg calcium, 78 mg phosphorus, 8 mg iron, 0.6 mg niacin, 78 mg vitamin C, 6050 μg beta carotene. The leaves contain between 7 - 12% oxalate. Flowers are eaten as raw with a nice acid flavour and a pleasant addition in the salad.

A slimy substance collects in the mouth when the leaves are chewed. This is used by magicians to protect the mouth when they eat glass. Yellow, orange and red to brown dyes are obtained from the flowers. The boiled whole plant yields a yellow dye.

BOTANY

Its trailing stems tend to root at the nodes, and there are small, squarish stipules next to the leaves. The leaves are compound, made up of two or more discrete leaflets, alternate or whorled, entire. Flower 4-8 mm, 1-6 in number, yellow, radially symmetrical. There are five sepals and petals, separate or tubular, stamens-10, anther is attached by its base to the filament, carpels fused, 5-carpelled, styles-5. Fruits are capsule, hairy, 8-25 mm long, 5-loculed,dry and splits when ripe.

Chemical Constituents

The plant methanolic and ethanolic extracts illustrate the presence of carbohydrate, glycosides, phytosterols, phenolic compounds, flavanoids, proteins (12.5%), amino acids and volatile oil. Leaves contain carbohydrate (1.36%), fatty acids (13.2%: Palmitic acid 1.8%, mixture of oleic, linoleic and lenolenic acid 3.8%), protein (12.5%), fiber (6.2%), tannin (0.62%), moisture (92.3%). Whole plant extract is assumed to be full of phytoestrogens. *Oxalis corniculata* is recognized with a number of compounds that are characterize by nuclear magnetic resonance, infrared, and mass spectrometry as (i) Oc-1, a mixture of saturated fatty acids C24 to C28; (ii) Oc-2, a mixture of long-chain alcohols C18 to C28; and (iii) Oc-3, a single compound that is a galacto-glycerolipid.

Agrotechniques

Amrul can grow very easily in humus rich sandy soil with warm climate. The tubers of 5-10 cm are buried in soil with 20 cm X 20 cm spacing. The plant flowers and fruits all the year round, but especially during September-October. Manuring is done mostly with compost. Weeding and mulching is necessary as and when required. Watering during the first month is essential and twice a week afterwards.

2. Common name: Ol, Elephant's Foot Yam
Scientific name: *Amorphophallus campanulatus*

Family: Araceae

INTRODUCTION

Elephant foot yam is botanically equated to *Amorphophallus campanulatus* (Araceae). It is a tropical tuber crop cultivated for most parts in Africa, South Asia, Southeast Asia and tropical Pacific islands. It is originated from Island Southeast Asia and spreads westwards into Thailand and India.

HABITAT

Primary habitat is secondary forests and dappled shade areas. It is terrestrial and abundantly grown in humus-rich places accumulated on rocks.

MEDICINAL PROPERTIES

The corms are dry, acrid, pungent; increase both appetite and taste; digestive, anthelmintic and aphrodisiac; useful in vitiated conditions of *vata* and *kapha*, elephantiasis, inflammations, haemorrhoids, haemorrhages, abdominal pain, asthma, piles, dysentery, splenopathy, amenorrhoea, seminal weakness, fatigue, anaemia and general debility. They are carminative, restorative, stomachic and tonic. After drying, they are used in the treatment of piles and dysentery.The fresh corms are also applied externally as an irritant to treat acute rheumatism. Corm contains betulinic acid, β- sitisterol, stigmasterol, triacotane, lupeol, and β-sitosterol palmitate. In addition, glucose, galactose, calcium oxalate, rhamnose and xylose are also present. The corms contain a diterpenoid (salviasperanol and amblyone) and a triterpenoids expressing antibacterial, antifungal and cytotoxic activities. Amblyone showed to have good antibacterial activity and moderate cytotoxic activity.3, 5-diacetylambulin, a flavonoid isolated from *A. campanulatus* showed antibacterial activities against four gram positive and six gram-negative bacteria. Research ensured potent hepatoprotective action against carbon tetrachloride-induced hepatic damage. The probable mechanism of antioxidant activity may be due to the free radical scavenging potential from the flavonoids in the extracts. Superoxide dismutase, catalase and glutathione peroxidase signifies hepatoprotective and antioxidant properties.

Other Uses

The corms are acrid, astringent, thermogenic, irritant useful in vitiated condition of *Vata* and *Kapha*. The stem is cut, and the inside of the stem is eaten raw in the treatment of snakebites. Rhizomes are preferably cooked, as acrid when raw and cause perioral burning and itching. After thorough cooking, petioles of young unexpanded leaves become edible. In time of scarcity, corms are sometimes eaten. In India, corms are used in curries and pickles.

BOTANY

Elephant's foot yam is a perennial, stemless herb. Corm is depressed-globose, up to 30 centimeters in diameter, flowering occurs before leafing every year from the previous year's corm. Stem-like structure, which bears the lamina, is merely the petiole, 1 meter or more in height, radially developed from the corm. Leaves are usually solitary, the blades are up to 1 meter in diameter, trisected, the segments dichotomous, the ultimate ones pinnately divided into oblong to oblong-obovate, acuminate lobes. Spathe is sessile, broadly campanulate, dull-purplish, the margins somewhat spreading or recurved, waved and crenulate, up to 30 centimeters in diameter. Spadix (a spike of flowers contained in the spathe) is hardly longer than the spathe, the appendage ovoid, variously sulcate or depressed, up to 15 centimeters long, are malodorous when flowering.

CHEMICAL CONSTITUENTS

Corm analysis resulted in 74% moisture; 0.73% ash; 5.1% protein; 18% carbohydrate, 0.61% crude fiber, giving 1,000 calories per kg. The crop yields amblyone, a tritepenoid with antibacterial and cytotoxic activities. Phytochemical screening of various tuber extracts yield alkaloids, carbohydrates, coumarins, steroids, saponins. Tubers produce an active diastatic enzyme- amylase, betulinic acid, B-sitosterol, stigmasterol, B-sitosterol palmitate, lupeol, triacontane, amino acids, carbohydrates, saponin, thiamine, riboflavin, niacin and carotene. Flavonoidal content is 46.33 mg/g (ME) and 36.88 mg/g (AC). Total phenolic contents are 12.67 mg/g (ME) and 6.25 mg/g (AE).

Agrotechniques

A temperature in the range of 25 - 35°C is required with an annual rainfall of 1,000 - 1,500 mm. Corm development is promoted during dry periods. The best crops are produced on deep, well-drained, humus-rich, fertile, alluvial soils that are slightly acidic. Heavy clay soils are not suitable for the crop. Plants require some shade, especially when young. The plants are usually grown on a three year cycle. The seed tubers (250-500 g size) are generally sown during February- March in pits of dimension 60 cm X 60 cm X 60 cm with 2 kg FYM and 50 g Sufala 15:15:15 mixed well with the soil. An interesting technique used in Indonesia is to dig up the tuber after one year and then replant it upside down. This stimulates the lateral buds into growth and increases the overall size of the tuber. When the plants are 30-45 days old, the basal soil is loosened and mixed with required fertilizer and water. Yields of upto 20 tonnes per hectare can be achieved. There are many cultivated forms, some of which have a lower content of calcium oxalate crystals and less or no alkaloids. The plant produces one or two leaves and one short flowering stem annually. The flowers have a foetid smell to attract carrion flies and midges favouring pollination. This smell disappears once the flowers have been pollinated. It also has the ability to heat the flowering spadix as the pollen becomes ready for fertilization. This heat greatly increases the strength of the aroma released by the plant, thereby attracting more pollinating insects. It can also have the effect of making the insects more active, thus increasing the level of fertilization.

3. Common name: Kalmi, Water spinach, Water morning glory
Scientific name: *Ipomoea aquatica*
Family: Convolvulaceae

INTRODUCTION

Water morning glory is a semi-aquatic tropical plant grown as a leaf vegetable especially in Asian markets. It is harvested from the wild but also commonly cultivated, especially in Asia. Its precise natural distribution is unknown due to extensive cultivation, with the species found throughout the tropical and subtropical regions of the world. Water Morning Glory grows in water or on moist soil. It is most commonly grown in East and Southeast Asia. It flourishes naturally in waterways and does not require much care. The plant has traditional medicinal uses. It is used extensively in Malay and Chinese cuisine.

HABITAT

Ipomoea aquatica is found in fresh water aquatic habitats. *I. aquatica* has been found primarily in canals and ditches but also invades lakes. It grows well in moist soil or in still to flowing waters.

I. aquatica dwells in muddy stream banks, freshwater pond and lakes, and marshes as well. It is confined to the tropics and subtropics, susceptible to frosts and does not grow well when temperatures are below 23.9^{0}C. As a consequence of its prolific growth habit this species can invade moist cultivated areas (such as rice and sugar cane fields) and other areas with varying water levels.

Medicinal Properties

In Ayurveda, leaf extracts are used against jaundice and nervous debility. In indigenous medicine in Sri Lanka, it is supposed to have insulin-like properties. The young shoots are mildly laxative and are used by diabetic patients. A decoction of the leaves is used to treat coughs. The fried leaves are eaten to cool down a fever. The crushed leaves are applied as a poultice on sores and boils. A paste made from the buds is used to treat ringworm. The roots are laxative, tonic and antidote. They can be used for healing of opium or arsenic poisoning, and also to counter the effects of drinking unhealthy water. A decoction of the roots is used as a wash against haemorrhoids. The plant has shown oral hypoglycaemic activity in tests with diabetic humans. Plant extracts of Ipomoea have moderate anti-cancer activity and can inhibit cancer cell growth.

Other Uses

Leaves and young shoots are generally used in raw or cooked forms. Roots are occasionally cooked and eaten. The tasty leaves are produced all the year round. The youngest shoot tips can be added to salads and older leaves are cooked. The leaves can be stir-fried, steamed, boiled for a few minutes or lightly fried in oil and eaten in various dishes. They are often mixed with hot peppers and garlic and prepared with a savoury dish. Very nutritious, the leaves are a good source of protein as well as vitamin A, iron, calcium and phosphorus.

BOTANY

Swamp morning glory is a fast-growing, annual or perennial plant with prostrate stems 2 - 3 metres long that sprawl over the ground, float in water or twine into other plants for support. The plant roots freely at the nodes and can form large clumps of growth. It is a smooth, widely spreading vine, with the stems trailing on mud or floating on water. Leaves are oblong-ovate, 7 to 14 centimeters long, with a pointed tip and heart-shaped or arrow-shaped base, long petioled, the margins entire or angular, and sublobed. Peduncles are erect, 2.5 to 5 centimeters long, with 1 or 2 flowers, borne in the axils of the leaves. Sepals are green, oblong, about 8 millimeters long. Corolla is narrowly bell-shaped, about 5 centimeters long, and purplish; limb nearly white or pale pink purple, about 5 centimeters in diameter, the tube deeper purple inside. Capsules are smooth and ovoid, about 1 centimeter long.

CHEMICAL CONSTITUENTS

Plant contains protein 1.6%, fat 0.2%, and vitamins C and B2. Some major compounds from the plants are: nortropane alkaloids, calystegines; phenolic compounds N-cis-Feruloyltyramine, N-trans-feruloyltyramine, 3α,7β-O-D-di glycopyranosyl-dihydroquercetin and Isochlorogenic acid a, b and c. Compositional analysis of leaves yield moisture 90%, protein 3%, fiber 0.9%, fat 0.4%, carbohydrate 4.3%, mineral matter 2%, nicotinic acid 0.6 mg/100g, riboflavin 120 mg/100g, vitamin C 137 mg/100g, vitamin E 11 mg/100g, and ash 1.4%. Nutrient analysis of leaves (based on % dry weight) show the leaves with 3% protein, 4.5% total carbohydrate, 0.3% lipids, 78% moisture, 1.6% ash, 1.4% fiber and gross energy value of 141.4 Kj/g. Leaves yield high amounts of essential amino acids (per 100g) 4765 mg: leucine 1365 mg/100g, tyrosine+phenylalanine 1124 mg/100g, lysine 682 mg, and threonine 606 mg. Minerals (per 100 g) are potassium 444 mg, calcium 163, sodium 159.8 mg, phosphorus 86, magnesium 52 mg, copper 5.3 mg, zinc 4.1 mg, iron 3.2 mg, and manganese 2.3 mg.

AGROTECHNIQUES

Swamp morning glory is a plant of lowland tropics with elevations up to 1,000 metres. It grows best in areas where annual daytime temperatures are within the range 15 - 35°c, but can tolerate 10 - 40°C. It prefers a mean annual rainfall in the range 2,000 - 2,500 mm, but tolerates 700 - 4,200 mm. Plants grow well in full sun. A very easy plant to grow, it succeeds in bathtubs, ponds and still water. It succeeds in a wide range of soils, including heavy clays. Best leaf production comes from plants grown in soils rich in organic matter. It prefers a pH in the range 5 - 7, tolerating 4.3 - 7.5. The plant is prevalent as a marshy weed in all tropical and many subtropical lowland areas. It is a declared aquatic or terrestrial noxious weed in the south-eastern United States. A very quick-growing plant, it starts developing lateral branches from cotyledonary buds 2 - 3 weeks after sowing. Subsequently, the main axis and both laterals produce about one leaf each on every 2 - 3 days. Cultivars selected for once-over harvest or uprooting have retarded branching or almost no branching. Their harvest takes place 21 - 30 days after sowing. Vegetatively propagated plants or wild plants can be harvested in about one month after plant establishment. Flowering commences after 2 - 5 months, but the plant continues to form new leaves and branches. Being a quantitative short-day plant, early flowering of this plant is induced by short days of less than 12 hours. Flowering is also stimulated by drought.

4. Common name: Chalkumro, Ash gourd, White gourd, Wax gourd
Scientific name: *Benincasa hispida*
Family: Cucurbitaceae

Introduction

Benincasa hispida is native to South and Southeast Asia. It is abundant ly grown in Asia, including Java, Japan and Sri Lanka. White gourd is basically a vine grown for its very large fruit, eaten as a vegetable. The plant can produce such enormous fruits that it either needs to be really well supported when climbing or be grown as a ground-cover crop. The plant is a rather important market vegetable in subtropical and tropical Asia. It is frequently cultivated, especially in the Old World tropics, for its edible fruit. It is also often used medicinally, especially in China.

Habitat

Not known in a truly wild situation. It is mainly found in riverbeds or furrow.

MEDICINAL PROPERTIES

It is a gentle and safe remedy for a number of complaints, especially as an effective tapeworm remover for children and pregnant women for whom stronger acting and toxic remedies are unsuitable. The leaves are applied externally to burns. The sap of the plant and the pulp of the fruit can also be used. The fruit pulp is used as a decoction to relieve intestinal inflammation. The seeds are mildly diuretic and vermifuge. The seed is high in zinc and has been used successfully in the early stages of prostate problems and to treat hypertrophy of the prostate. The diuretic action has been used in the treatment of nephritis and other problems of the urinary system.

Young ashgourd removes gastric, old ashgourd removes cough. Ripe ashgourd removes urinary and coprolite disorder. The juice of ashgourd is effective in blood gastric. The additional substance of dry honey is also effective in headache. The juice is accepted with honey in pluricy. Ripe ashgourd with cow milk is also effective in heart growth. The mixture of hot milk and ripe ashgourd is effective in conspiracy. The juice is also effective in tuberculosis. The kernel of ashgourd in hot condition is used in bellyache. Juice of ashgourd is effective against memory lossand the kernel with water is also effective in vermes disease.

OTHER USES

Fruits are consumed either raw or cooked and mainly used as vegetable. They are also used in pickles, curries, preserves and candied in sugar.The fruit can be eaten either when it is young or fully mature. It has a juicy texture with a mild flavour, the flavour being somewhat stronger in younger fruits. Because of its waxy coating, it can be stored for several months, sometimes as long as a year. Mature fruits can vary in weight from 2 - 50 kg. A nutritional analysis is available. Young leaves and flower buds are steamed and eaten as a vegetable, or are added as a flavouring to soups.

Seeds are rich in oil and protein. A wax that coats the fruit is used to make candles. The roots have considerable resistance to soil-borne diseases and they are sometimes used as a rootstock for melons and other cucurbits.

Botany

It is a robust, annual climbing plant producing stems up to 6 metres long that scramble over the ground or climb into other plants, supporting themselves by means of slender, rarely simple tendrils. The stem is much branched. Leaf stalks are long and hairy. Leaves are roundish, kidney-shaped, base deeply heart-shaped. Upper surface is rough, lower surface shortly bristly, blade 10-25 cm long and as much broad, 5-7-lobed, lobes ovate-triangular, margin sinuate or toothed. The fruit is fuzzy when young. By maturity, the fruit loses its hairs and develops a waxy coating, giving rise to the name wax gourd, and providing a long shelf life. Male flower is 5-15 cm long, female flowers 2-4 cm long. Calyx tube is 10-15 mm long, densely hairy, lobes lanceolate, acute, 6-12 mm long. Petals are spreading, blunt, but ending in a short point, 3-5 x 2-4 cm. Filaments of the stamens are inflated and hairy at the base, anthers trilobed.

Chemical Constituents

Phytochemical analysis confirmed presence of major constituents like volatile oils, flavonoids, glycosides, saponins, proteins, vitamins, minerals, β-sitosterin and uronic acid. The presence of sterols such as lupeol, lupeol acetate and beta-sitosterol; triterpenes like alnusenol, multiflorenol, iso multiflorenol and flavones like iso-vitexin has been reported.. *B. hispida* may contain isovanillin. The chemical name of isovanillin is 3-hydroxy-4-methoxybenzaldehyde.

AGROTECHNIQUES

Benincasa hispida is a plant of the tropics, and is also cultivated in the subtropics. It can be grown at elevations up to 1,500 metres. It grows best in areas where annual daytime temperatures are within the range 24-30°C. It prefers a mean annual rainfall in the range 400 – 800 mm. Excessive rain may reduce yields, but the water requirements are still fairly high especially in the early stages of development. Essentially it requires a warm sunny position in a rich well-drained soil and plenty of moisture in the growing season, though high rainfall can be detrimental. Established plants are reasonably drought tolerant. They prefer a pH in the range 5.6 - 6.8. The development of female flowers is stimulated by short day lengths and relatively cool weather conditions, while male flower production is encouraged by high temperatures and long days.

The fruits can be harvested about 3 months after sowing. Plants take up to 5 months from seed to produce a mature crop, though the fruits can be eaten when immature. It can be picked as early as one week after fertilization. Yields of 20 tonnes per hectare have been achieved in northern India, with individual fruits weighing from 5 - 45 Kg at maturity. The average yield in Taiwan is 40 tonnes per hectare, with an optimum of 50 tonnes. There are many named varieties. One group, sometimes classified as *B. hispida* chieh-gua, is known as the hairy melon or jointed gourd. This form is grown for its immature fruit in much the same way as courgettes are used. Mature fruits of this form do not develop a waxy coating.

5. Common name: Kata note, Spiny amaranth, Thorny amaranth
Scientific name: *Amaranthus spinosus*
Family: Amaranthaceae

Introduction

Amaranthus spinosus is an erect, much-branched annual plant growing up to 100 cm tall. The plant is harvested from the wild for local use as a source of food and medicines. The leaves are sometimes sold as a vegetable in local markets.

Habitat

Natural habitats are roadsides, waste places.

Medicinal Properties

The seed is used as a poultice for broken bones. The plant is astringent, diaphoretic, diuretic, emollient, febrifuge and galactogogue. Externally, it is used to treat ulcerated mouths, vaginal discharges, nosebleeds and wounds. The plant can be used fresh or it can also be harvested when coming into flower and dried for later use. The root is emmenagogue and galactogogue. A paste of the root is used in the treatment of menorrhagia, gonorrhoea, eczema and colic. It helps to remove pus from boils. In Nepal, the juice of the root is used to treat fevers, urinary troubles, diarrhoea and dysentery. It is also used, often combined with the root juice of *Dichrophela integra* and *Rubus ellipticus*, to treat stomach disorders and, on its own, to treat indigestion and vomiting that occur after eating unusual foods.

OTHER USES

Leaves and stems are consumed raw or cooked as spinach. If older leaves and stems are used the spines must be removed. With a unique bitter flavour, it is usually eaten in small quantities only as a substitute when no other vegetables are available. Seeds are very small, about 1mm in diameter, but easy to harvest and very nutritious. The seed can be cooked whole, becomes very much gelatinous. The ash of the burnt plant is used as a salt. It is also added to the water as a tenderizer when cooking rough vegetables such as cowpea leaves and pigeon peas. A red pigment obtained from the plant is used as a colouring in foods.

BOTANY

It is an annual, erect, monoecious herb, up to 100–130 cm tall, much branched; stem terete or obtusely angular, glabrous or slightly pubescent, green or variably suffused with purple. Leaves alternate, simple; stipules absent; petiole approximately as long as leaf-blade; blade ovate-lanceolate to rhomboid, 3.5–11 cm × 1–4.5 cm, acute and often slightly decurrent at base, obtuse, rounded or slightly retuse and often short mucronate at apex, entire, glabrous or slightly pubescent on veins when young. Inflorescence consisting of dense clusters, lowers ones axillary, higher ones often collected in an axillary and terminal spike which is often branched in its lower part; axillary clusters usually armed with 1–3 very sharp spines up to 2 cm long.

Chemical Constituents

The dried leaves contain (per 100g) 267 - 276 calories, 20 - 34.4% protein, 2 - 4.5% fat, 45 - 54% carbohydrate, 9.8 - 10.4% fibre, 16.6 - 24% ash, 1795 – 5333 mg calcium, 333 – 460 mg phosphorus, 13.5 - 152.7 mg iron, 13 – 37 mg sodium, 337 – 3528 mg potassium, 27.9 - 40.8 mg β-carotene equivalent, 0.06 mg thiamine, 2.02 mg riboflavin, 7.7 - 8.6 mg niacin and 503 mg ascorbic acid.

Agrotechniques

It usually prefers a well-drained fertile soil in a sunny situation. The crop grows best in a moist soil, but is able to tolerate fairly dry soils and also quite wet, but not waterlogged, ones. The plant produces seed prolifically, with almost 250,000 seeds being recorded from one plant. It has spread as a weed of disturbed ground from its original range in the Americas to many other areas of the world. Most of the members of this genus photosynthesize by a more efficient method than other plants, called the C-4 carbon-fixation pathway. This process is particularly efficient at high temperatures, in bright sunlight and under dry conditions. Plants should not be given inorganic fertilizers. After thinning, mostly organic fertilizer should be applied with proper irrigation. It can be cultivated with less care and management. Plants are predominantly susceptible to attacks by leaf-chewing insects.

6. Common name: Kochu, Taro, Cocoyam, Dasheen
Scientific name: *Colocasia esculenta* Schot/ *Colocasia antiguor*
Family: Araceae

Introduction

Prickly Amaranth is an annual herb with red tinged erect stems. Different types of colocasia (kochu) are available viz., maan kochu, mukhi kochu, panchamukhi kochu, sola kochu, padma kochu, sobuj kochu, lal kochu, kalo kochu, dudh kochu, guri kochu, loti kochu, etc. It is grown in normal places. Root and stem are edible. Leaves are also popular in edible purpose.

Habitat

C. esculanta is an erect perennial plant, grows upto 1.5 m height. Cut leaves produce transparent latex. It rarely flowers. It is an upright, erect annual herb, many branched and can grow up to 1 m high.

Medicinal Properties

The seed is used as a poultice for broken bones. The plant is astringent, diaphoretic, diuretic, emollient, febrifuge and galactagogue. It is used in the treatment of snake bites, internal bleeding, diarrhoea and excessive menstruation. It is applied externally to treat ulcerated mouths, vaginal discharges, nosebleeds and a range of skin problems including wounds, eczema, boils and burns. The plant can be used fresh or it can also be harvested when coming into flower and dried for later use. The ash of the plant, combined with water, is used to wash sores. The sap is used as an eye wash to treat ophthalmia and convulsions in children.

BOTANY

C. esculenta is a perennial, herbaceous rodless herb, 1.5 m tall, with thick shoots from a large corm; slender stolons also often produced, along with offshoot corms. Roots are large, underground tuberous rhizome. Leaves have a long thick and fleshy petiole, longer than the lamina, up to 85 cm long, sheathing at the base. The lamina is large, triangular or arrow head, completely smooth surfaced and very much hydrophobic. Leaf blades to 60 cm long and 50 cm (wide, arrowhead shaped, with upper surface dark green and velvety; leaves peltate (stalked from back of blade); petioles large, succulent, often purplish near top. Inflorescence is on a fleshy stalk shorter than leaf petioles; part of fleshy stalk enveloped by a long yellow bract (spathe). Flowers are tiny, densely crowded on upper part of fleshy stalk, with female flowers below and male flowers above. Flowers are pale yellow to yellow-orange, arranged in spike inside a spathaceous envelope. Fruit, a small berry, remains in clusters on the fleshy stalk.

CHEMICAL CONSTITUENTS

Mainly leaves contain calcium oxalate, fibers, minerals (calcium, phosphorus), starch, and vitamin A, B, C, etc. Phytochemically, these also contain flavones, apigenin, luteolin, and anthocyanins.

Agrotechniques

It prefers a well-drained fertile soil and a sunny situation. Grows best in a moist soil, but is able to tolerate fairly dry soils. It doesn't like waterlogged ones. The plant produces seed prolifically, with almost 250,000 seeds being recorded from one plant. Seeds are sown in situ. Germination is usually rapid and good if the soil is warm, but poor germination rates are experienced in cool or cold soils. A drop in temperature overnight aids in germination. Cuttings of growing plants root easily.

7. Common name: Khesari, Grass pea, Indian pea, Chickling vetch
Scientific name: *Lathyrus sativus*
Family: Fabaceae

Introduction

Lathyrus sativus was one of the earliest cultivated food crops in Europe with a history of cultivation dating back perhaps 8,000 years. It is a minor crop found especially in the Indian subcontinent, in the Middle East and North Africa, and also to a lesser extent in America, Australia and South Africa. In tropical Africa it is mainly grown in Ethiopia, but is also found in Sudan, Eritrea, Kenya, Tanzania, Angola and Mauritius. The plant is occasionally cultivated for its edible seed and as a green manure.

HABITAT

It is normally found as a weed of cultivated land. The plant is not known in a truly wild state.

MEDICINAL PROPERTIES

Used to protect from constipation, body pain, ricket etc.The oil from the seeds is a powerful and dangerous cathartic. It possesses many pharmacological effects including antioxidant, nervous, antidiabetic, analgesic, antipyretic and cardioprotective effects.

OTHER USES

The immature seed can be eaten like green peas. The mature seed is eaten cooked. It needs to be soaked and well cooked before being eaten. The seed can be eaten in a wide variety of ways:

- The seeds are often simply boiled and eaten
- They are split and then made into a sauce and eaten with unleavened bread in Ethiopia
- They are made into dahl and rotis in India
- They can also be ground into a powder then mixed with wheat in a ratio of one part chickling pea to three parts of wheat flour to make a protein-enhanced bread.

Young seed pods - cooked and eaten as a vegetable. They can be boiled, salted and consumed as a snack. Young shoots – cooked.

Botany

Lathyrus sativus is a much-branched, annual plant growing from a long but thin, well-developed taproot; it can grow 30 - 50cm tall, occasionally to 170cm. The sub erect stems often scramble into the surrounding vegetation, supporting themselves by means of tendrils.

Chemical Constituents

It contains starch, cane sugar, leguminvicilin, legumelin, fixed oil, gum resin, oleoresin, β-carotene, proteins and amino acids.

Agrotechniques

Chickling pea is a plant of temperate to subtropical regions, though it can also be cultivated in the tropics, especially at higher elevations (in Ethiopia, it is grown at elevations up to 2,700 metres). It grows well in the subtropics as a winter crop. An easily grown plant, it does best in a sunny position but can be grown on a wide range of soil types, including poor soils and heavy clays. It tolerates waterlogging and moderate salinity. Flowering time is 6 - 17 weeks after sowing.

It is best to only remove the aerial parts of the plant, leaving the roots in the ground to decay and release their nitrogen. In order to avoid shattering the seedpods and scattering the seed, harvesting of the seeds should be carried out when the leaves turn yellow and pods turn grey. The plants are pulled out by hand or cut with a sickle near the base. They are then stacked and allowed to dry for 7 - 8 days in the field or on the threshing floor. The average seed yield is 350 - 700 Kg per hectare, whereas, it may be more than that in some other countries.

Grass pea is grown as a sole crop or intercropped with plants such as barley, linseed or chickpea. In many countries it is produced in rice-based cropping systems before the rice crop or alternately with a rice crop. In India it is often grown as a relay crop: it is broadcast into a standing rice crop about two weeks before the rice harvest and left to grow on the residual moisture. The plant has an extensive root system and fixes atmospheric nitrogen through bacteria that live on the roots. It makes a good soil-enriching green manure crop or can be planted for erosion control.

8. Common name: Gadal pata, Stinkvine, Skunkvine, Chinese fever vine

Scientific name: *Paederia foetida*

Family: Rubiaceae

INTRODUCTION

It is a fast-growing, slender, perennial, climbing plant. The whole plant has a foetid aroma when rubbed. It is a popular medicinal plant in the Orient. It is often grown as an ornamental.

Habitat

Found in thickets and woodland, also along secondary evergreen to deciduous forest and clearings in primary forest. It can grow in montane vegetation up to 3,000 m, on steep, forested slopes, or on sandy or rocky sea coasts.

Medicinal Properties

Considerable research has been carried out into the medicinal benefits of the plants, with several active compounds having been recorded. The leaves are widely used in Asia and South-East Asia, where they are predominantly exquisite for treating digestive problems.

A number of iridoid glucosides such as asperuloside, scandoside and paederoside have been isolated from the aerial parts of the plant. These glucosides have displayed antitumor activity with paederoside displaying the highest degree of antitumor-promoting activity. Scandoside has been shown to promote the growth of lettuce seedlings. Paederoside showed a similar inhibitory activity to that of asperuloside and daecetyl-asperulosidic acid in a plant growth inhibition test and in an antimicroorganism activity test. Decoctions of the plant showed significant anti-inflammatory action against arthritis. The decoction also exhibited marked activity against degenerative osteo-arthritis.The water soluble fraction of the aerial parts displayed anti-inflammatory activity in oedema. It failed to exhibit any analgesic or antipyretic action and showed no ulcerogenic potential. An ethanolic extract of the leaves has shown anticancer activity against human epidermoid carcinoma of the nasopharynx.

The leaves are anodyne, antirheumatic, antivinous, astringent, carminative, depurative, diuretic, restorative and vermifuge. They are commonly used for the treatment of intestinal complaints such as abdominal pain, colic, cramps, flatulence and dysentery; and are also used for treating rheumatism and gout. The leaves are also used to treat infertility and paralysis. The leaves and stems are diuretic. They can treat intestinal problems.

OTHER USES

Leaves are consumed either as raw or cooked. Although they have a strong flavour, the leaves are sometimes mixed with grated coconut and spices, and then eaten raw as a side-dish with rice. The minced leaves are steamed and eaten; they can be added to soups; or are mixed with various vegetables and spices, then wrapped in a banana leaf and cooked over a fire. Any offensive smell disappears as the leaves are cooked. The leaves are valued more for their medicinal virtues than for their flavour. In Taiwan, the sweet stem juice is sucked.

The stem yields a strong and silky fibre, but it is not commercially exploited. An ethanolic extract of the leaves and stem has been shown to be significantly toxic against the aphid *Myzus persicae* infesting cabbage in India.

BOTANY

As its common name suggests, stinkvine is a bad-smelling climber. It produces slender stems 1.5 - 7 metres long that may reach upto 9 m. Stems are woody only at the base. Growth form and vine appearance are variable. It may grow prostrate or twine on other vegetation or supports. Ground-creeping vines form adventitious roots at the nodes. Leaves are also found in whorls of three. Petioles commonly measure up to 6 cm long, and leaf blades are typically 5-14 cm long and up to 5 cm wide.

Chemical Constituents

Paederia foetida includes paederolone, paederone, β-sitosterol, paederoside, glucosides, iridoid, asperuloside and various alkaloiuds etc. The plant also contains friedelin, campesterol, ursolic acid, hentriacontane, hentriacontanol, ceryl alcohol, palmitic acid and methyl mercaptan, ellagic acids, terpenoids and essential oil etc. The bruised aerial parts of the plant have the fetid odour of indole (methyl-mercaptan). Damage to the tissue releases an enzyme which splits off this sulphur-containing group from paederoside, and is therefore responsible for the unpleasant odour. The most abundant sulphur-containing compound was dimethyl disulphide. The presence of alkaloids and an essential oil have also been recorded. Monoterpenes such as linalool constitute the major components in the oil.

Agrotechniques

The plant is prevalent in a range of environments from the warm temperate zone to the tropics. It is tolerant of frost, at least partially. It can succeed in any fertile soil so long as it is well drained. Plants have a wide ranging adaptability to different levels of light, types of soil, and salt conditions. It normally equires a sunny sheltered position.

The plant has escaped from cultivation and become naturalised in many areas. It has become a serious weed of young sugarcane. The vines climb over shrubs and trees, weighing them down and impeding regeneration. They also invade pastureland and are troublesome along roads and on power lines. The seeds may be dispersed by birds and the plant is also spread by the transport of rooted fragments. The plant can be found flowering and fruiting throughout the year in tropical and subtropical conditions. In other localities, it flowers during the rainy season and fruits early in the dry season. Individual flowers are short-lived, open early in the morning and fall off after 2 days; entire inflorescences, however, bear flowers for a long period of time.

9. Common name: Gima, Bitter cumin, Bitter leaf, Indian chickweed
Scientific name: *Glinus oppositifolious*
Family: Molluginaceae

INTRODUCTION

Glinus oppositifolius is an erect or almost prostrate, much-branched annual herb with a strong taproot and stems that can be 10 – 60 cm long. It is occasionally cultivated in Thailand and the Philippines for use as a vegetable and is often sold in local markets. Flowering and fruiting almost year-round. $2n = 36$.

HABITAT

Prefers dry regions, grows from the plains up to 100 metres, exceptionally to 275 metres, in seasonally swampy localities, in dried-up pools and ditches and on rice-fields. Sometimes found in sandy localities near the sea.

MEDICINAL PROPERTIES

The plant is said to promote digestion and salivation. It is used as a treatment for bowel complaints and syphilitic affections.

Anti-Diarrhoeic Activity

Methanol extracts and loperamide at the dosages of 500 mg/kg and 3 mg/kg respectively, were reported to treat diarrhoeic rats and the total number of faeces and the number of diarrhoeic faeces were both reduced after treatment.

Anti-Depressant and Anxiolysis Activity

The crude methanol extract of *Glinus oppositifolius* has profound effects on the normal brain functioning of mice. Oral administration of the ethanolic extracts induced anxiolytic effects. Ethanol extract was found to reduce locomotion of the animals both in hole cross and in open field tests significantly. Diazepam was also seen to exhibit the same effects as ethanol extract in all the experiments. Ethanol extract also showed sedative effects on mice, which was proved by oral administration of the crude extract, where it enhanced the sleeping duration in a dose dependent manner.

Anti-Inflammation and Analgesic Activity

Water and ethanol extracts of *Glinus oppositifolius* exhibited anti-inflammatory activity, and the extracts showed high potency in inflammatory paw oedema assay and constrictive abdominal response assays. The results exhibited that the ethanol extract was more potent in its anti-inflammatory capacity than the water extract. The extracts were reported to have a characteristic profile of nonsteroidal anti-inflammatory drugs (NSAID) in their activity.

Anti-Diabetogenic Activity

Alloxan was used to induce diabetes in mice and *G. oppositifolius* was administered which exhibited decrease in blood glucose level compared with the control group. Doses of 200 mg/kg and 400 mg/kg showed a concentration-dependent glucose level lowering activity when compared to the control. The results were found to be statistically significant ($p<0.05$-0.001).

OTHER USES

Leaves are cooked and used as a vegetable with slight bitterish flavour.

BOTANY

Jima is a diffuse or prostrate, annual, much branched, hairless herb. Leaves are in pseudo whorls of 3-6 or opposite; leaf blade spoon-shaped or elliptic, 1-2.5 cm × 3-6 mm, base attenuate, margin with sparse teeth, apex obtuse or acute. Leaf stalks are short. Flowers are axillary, usually 2-7 in a cyme, greenish white in colour around 5- 8 mm across. Perianth segments are white or tinged with pink. Tepals are 5, greenish white, yellowish, or milky-colored, oblong, 3-4 mm, margin membranous, 3-veined. Stamens are 3–5, often alternating with linear, bifid staminodes. Carpels are 3; styles 3. Seeds are numerous. Ovary is ellipsoid, 3- or 4-valved. Seeds are ovoid, curved, tubercled, chestnut-brown, subreniform, granulose; strophiole, 1/5 as long as seed, claviform.

Chemical Constituents

A huge number of chemical constituents have been reported from *Glinus oppositifolius*. The vast array of phyto-constituents present in *Glinus oppositifolius* can be broadly categorized into several groups- Different aromatic compounds are extracted from *Glinus oppositifolius*, viz., flavonoid glycosides, acylamino derivatives and nucleosides. Different triterpenoids and triterpenoidal saponins extracted from *Glinus oppositifolius*, Different sterols extracted from *Glinus oppositifolius*. The other aromatic compounds reported from *Glinus oppositifolius* are benzoic acid, 4-hydroxybenzoic acid, 4-hydroxybenzaldehyde, 4 hydroxy-acetophenone, methyl 4-hydroxybenzoate, anisic acid, vanillin, 4-hydroxy-3-methoxyacetophenone, acetosyringone, 4- hydroxy-3,5-dimethoxy benzaldehyde, 4-hydroxybenzyl alcohol, 2-(4- hydroxyphenyl ethanol, cinnamic acid, trans-ferulic acid. The flavonoid glycosides which are reported to be present in *Glinus oppositifolius* are Kaempferol 3-O-galacto pyranoside, isorhamnetin 3-O-β-D-xylopyranosyl-β-D-galactopyranoside, vitexin, and vicenin-2.

Agrotechniques

It is as such not generally cultivated, rather growing automatically near swampy areas, like near well, water pump, pond etc. It can be cultivated by sowing seed and from planting of cutting. Flowering occurs almost all the year round.

10. Common name: Chikni sak, Common knotweed, Small knotweed
Scientific name: *Polygonum plebejium*
Family: Polygonaceae

Introduction

Polygonum plebejium is a minor neglected annual plant popularly known as common knotweed or small knotweed.The plant is harvested from the wild for local use as a vegetable and medicine as well.

Habitat

It is gregarious by rivers, canals, drying out pools and harvested rice fields, all areas which have clayey soil and have water logging at least some parts of the year. It forms dense mats on ground. It is also found as a weed in fields.

Medicinal Properties

The crushed cooked aromatic seeds are suitable for eating as a remedy for bowel complaints. The roots are similarly used. The fresh root contains 11% tannin, and oxymethylanthraquinone has also been isolated. This is also reported to be used as a demulcent, pectoral, tonic and diuretic. Various researches have shown this plant possessing considerable antimicrobial, anti-inflammatory, hepatoprotective, hypotensive and anticoagulative properties. In Unani system of medicine, these are considered laxative and diuretic. It is also helpful to cure stomachache and pain in bladder. *P. viviparum* are considered a good tonic, styptic; and also used against problems for chest and lungs, piles, old diarrhoea, rhinitis, vomiting, biliousness, chronic bronchitis, wounds, gripping in the abdomen etc.

Other Uses

Leaves can be cooked with potatoes and groundnuts and consumed as a vegetable. The food is slimy, but is well liked because it has a good smell. In India it is used as a famine vegetable.

Botany

It is a prostrate annual herb, much-branched, glabrous; stems up to 35 cm long, scabrid, red-brown, with short internodes. Leaves alternate, simple; cylindrical, up to 3 mm long, often silvery-white, unevenly fringed; petiole very tiny; blade very minute, linear to obovate-elliptical, 1–2 cm × 2–5 mm, margin revolute, somewhat thick and leathery, dark green turning red. Inflorescence is a congested raceme with short branches, forming axillary, 1–5-flowered clusters. Flowers bisexual; perianth 2 mm long, greenish, with 4 lanceolate-elliptical lobes 1.5 mm long, the outer pair keeled, white to pale pink; stamens 5–8; ovary superior, 1-celled, styles 3, free, 3 mm long. Fruit is basically a trigonous nut up to 2 mm long, smooth and shiny black.

Chemical Constituents

Gallic acid, rutin, coumaric acid, and quercetin were found in the leaves of *P. minus*. Polyphenolic compounds, such as hydroxyanthraquinones, stilbenoids, and proanthocyanidins, are abundant in Polygoneae plants.

Agrotechniques

When growing in dry places the leaves are said to develop a bitter flavour. The plant is sometimes found as a weed in cultivated fields.

11. Common name: Chola, Bengal gram, Egyptian pea
Scientific name: *Cicer arietinum*
Family: Fabaceae

Introduction

A staple item of food in many areas of the world, chickpea has probably been in cultivation by humans for over 10,000 years. It is broadly cultivated in warm temperate and tropical areas for its edible seed.

HABITAT

The genus *Cicer* is spontaneous plant in the warm regions of the Old World, with particular presence in arid or semiarid climates. Most species are concentrated in central and western Asia.

MEDICINAL PROPERTIES

Cicer arietinum possesses aphrodisiac, estrogenic, antioxidant, antifungal, antidiabetic, anti-inflammatory, hypocholesterolaemic, antidiarrhoeal, anticonvulsant, hepatoprotective, anticancer, diuretic, and antinephrolithiasis activities. Seeds of *C. arietinum* are used against stomach disorders such as diarrhea, dyspepsia, indigestion, vomiting, and dysentery. Intake of seeds is helpful against constipation, cholesterol, and nervous system disorders. Uncooked seeds act as retrigerant, diuretic, and stimulant. Seeds are advocated to atherosclerosis patients as a preventive diet. It also aids in bronchitis, liver and spleen enlargement, and skin diseases. The seeds are used to control swelling of liver and spleen, chest problems, throat infection, foul mouth, and pyrexia.

OTHER USES

The fresh or dried seed is consumed either as raw or cooked in soups, stews etc. It has a fairly sweet flavour and a floury texture that is to a certain extent indicative of sweet chestnuts (*Castanea sativa*). The mature seed can also be sprouted and eaten raw. Parched seeds can be eaten as a snack. The seed can also be ground into a meal and used with cereal flours for making bread, cakes etc. The seed meal, mixed with sesame paste, yields the well-known appetizer 'hummus.' The seed is a good source of carbohydrates and protein. The roasted seed is a coffee substitute. Both the young seedpods and the young shoots are said to be edible with some cautions to avoid toxicity. A refreshing drink can also be made from seedpods.

Botany

Chickpea is an edible legume, and an annual herb. The plant grows to between 20 and 50 cm high and has small feathery leaves on either side of the stem. The plant produces a taproot reaching 1 - 2 metres deep, with secondary roots mostly spreading in the top 15 – 30 cm soil layer. One seedpod contains two or three peas. The flowers are white or sometimes reddish-blue.

Chemical Constituents

Some active compounds of *C. arietinum* are daidzein, matairesinol, formononetin, secoisolariciresinol, genistein, and biochanin A3. Plant comprises of carotenoids (α-carotene, β-carotene, lycopene, and lutein), carbohydrates (Manninotriose, stachyose, verbascose, and raffinose), vitamins (A, B, and C), minerals (Phosphorous, potassium, calcium, and folate) and protein.

Agrotechniques

Plants grow well in lowland tropical areas, up to an elevation of 1,200 metres. They are well-adapted to semi-arid areas with a rainfall of 650 - 1,000 mm per year and are intolerant of heavy rain. They often do not succeed in mild moist maritime climates since the seedpods are hairy and this holds moisture. The moisture then encourages fungal growth and the seed usually rots before it is fully mature. Daytime temperatures in the range 18 - 26°C, plus a night time drop of around 8°C, are optimum. Plants become more tolerant of high temperatures in the latter stages of growth.

It prefers a light well-drained fertile soil. Clay loams are generally suitable. It requires a hot sunny position. It can tolerate drought to some extent once established and also a pH in the range 6 to 9, succeeding in alkaline and saline conditions. Pods can be harvested 3 - 5 months after sowing. An annual yield of 500 - 2,000 kg of seeds has been obtained.

12. Common name: Tok palangsak, Flame flower
Scientific name: *Talinum portulacifolium*
Family: Portulacaceae

INTRODUCTION

Talinum portulacifolium is a perennial, herbaceous plant with grey, tender stems which sprouts from a thickened root that is occasionally creeping. The stems, can sometimes become softly woody and frequently grow up to 1 m long, seldom to 3 metres.The plant is gathered from the wild for local use as a food and medicine. It is sometimes cultivated in tropical Africa and India as a vegetable, and is also grown as an ornamental plant.

Habitat

It is very common in a wide range of habitats ranging from rocky locations to grassland, in open bush land and thickets, along water courses and as a weed in fields, at elevations from sea-level up to 2,200 metres.

Medicinal Properties

It is a very digestive and tasty leafy vegetable. It can protect rheumatism, coughing and bile problem. The leaves are used in the treatment of eye diseases. A decoction is used as a remedy for constipation. In Ethiopia, the roots are used against cough and gonorrhoea.

Other Uses

Leaves and young stems are consumed either raw or cooked. It is used as a vegetable or added to salads. The leafy stem parts are often cooked together with Bidens or Cleome leaves, or mixed with coconut milk or crushed groundnuts and eaten with the staple food. The leaves can also be stored dry for later use. In Tanzania the plant is credited with aphrodisiac properties and a leaf decoction is used as a remedy for constipation. The plant is an appetizing fodder for cattle and goats.

Botany

Erect annual subshrubs with grey, succulent, glabrous stems up to 1–3 m long, sprouting from a thickened root. Leaves alternate, simple, slightly fleshy; petiole 2–3 mm long; blade broadly obovate to oblong, 2–8 cm × 1–3.5 cm, base cuneate, apex rounded but apiculate, margin entire. Inflorescence is a terminal panicle with lateral cymes on a central axis up to 30 cm long. Flowers bisexual, regular, opening towards the evening, 2–2.5 cm in diameter; pedicel 1–2 cm long, recurving in fruit; sepals 2, ovate, 4–6X3 mm g, concave; petals 5, 9-12 X5-6 mm, obovate, usually magenta, sometimes red, pink or white; stamens 25, filaments 2-3.5 mm long, unequal, basally connate, anthers 1 mm long, oblong; ovary superior, style slender, 1.5 mm long, ending in a 3-branched stigma. Fruit is a globose capsule up to 7 mm in diameter, bright yellow, 3-valved, the valves falling separately, many-seeded. Seeds are lens-shaped, 1 mm in diameter, 35 in each capsule, glossy black, shining, smooth or papillose with concentric striations.

Chemical Constituents

It contains a number of chemical constituents viz., protein, starch, neutral detergent, fiber, ash and lipids. Generally, three important chemical constituents of fenugreek are significant, i.e.,

1) steroidal sapogenins
2) galactomannans and
3) isoleucine

These constituents have positioned fenugreek among the most universally renowned "nutraceutical" or health food products. Other foremost constituents of fenugreek are saponins, flavonoids and alkaloids. The bitter taste and specific smell of fenugreek is due to alkaloids and some other volatile compounds.

13. Common name: Data note, Green amaranth, Slender amaranth
Scientific name: *Amaranthus viridis*
Family: Amaranthaceae

INTRODUCTION

This plant was cultivated by the ancient Romans and Greeks for its edible leaves. Gradually, the plant has become a weed in several areas of the world outside its original range. It is classified as 'Invasive' in many Pacific Islands. The edible leaves and seeds are occasionally collected from the wild and used locally. The plant is often endorsed to grow as a weed and is seldom cultivated as a vegetable. The leaves are sold in local markets.

HABITAT

A cosmopolitan weed growing on waste ground

Medicinal Properties

Amaranthus viridis is used as a traditionalmedicinein the treatment of fever, pain, asthma, diabetes, dysentery, urinary disorders, liver disorders, eye disorders and venereal diseases. The plant also possesses anti-microbial properties.

The plant is an emollient and vermifuge. A decoction of the whole plant is used to prevent dysentery and inflammations, and also to decontaminate the blood. The root juice is used to treat inflammation during urination. It is also taken to treat constipation. The crushed root is useful against dysentery. The leaves are diuretic, febrifuge and purgative. The leaf sap acts as a vermifuge, being effective against filaria, as an emmenagogue and to relieve heart troubles. The leaves are used in poultices (fresh or as dried powder) to treat inflammations, boils and abscesses, gonorrhoea, orchitis and haemorrhoids. The leaf sap is used as an eye wash to treat eye infections.

Other Uses

Leaves are consumed raw or cooked as spinach. The leaves include about 3.88% protein, 1.1% fat, 9.38% carbohydrate, 3.2% ash, 323mg Ca, 8.3mg Fe. They are also very rich in Vitamins A, B1& C. Amaranth is also used as a cereal substitute in cakes, porridge etc. It is uncomplicated to harvest and incredibly nutritious. The seed can be cooked whole, and becomes very gelatinous. An edible dye is obtained from the seed capsules. Yellow and green dyes can also be obtained from the whole plant.

Botany

- *Amaranthus viridis* is a vigorous, annual plant producing erect to prostrate stems 6 – 90 cm long. The stems can be simple or branched, sometimes radiating from base and forming mats. Flowers have short bracts so that the inflorescence is not bristly, its spreading styles are separated by a 'U'-shaped cleft, and the sepals are obtuse.

- *Amaranthus cruentus* is the most photoperiod insensitive and widely-adapted grain species. Its flowers also have short bracts, but its styles are vertical and parallel.
- *Amaranthus hypochondriacus* has been adapted to temperate photoperiods by plant breeders and is not disease prone. Its flowers have bracts that roughly equal the height of the styles, making the inflorescence bristly. Its styles join with a V-shaped cleft.

CHEMICAL CONSTITUENTS

Amaranth grain contains phytochemicals that are not defined as nutrients and may be antinutrient factors, such as polyphenols, saponins, tannins, and oxalates. These compounds show antinutrient effect by cooking.

AGROTECHNIQUES

Often cultivated, the plant succeeds from the temperate zone through to the tropics. It grows best in areas where annual daytime temperatures are within the range 21 - 27°C. It can be killed by temperatures of 4°C or lower. It prefers a mean annual rainfall in the range 1,200 - 1,500 mm. It prefers a well-drained fertile soil in a sunny position. It requires a hot sheltered position and grows in a pH ranging 5.5 - 7, tolerating 4.3 to 7.5.

For stem it can be sown in June-July, for leaf it can be sown throughout the year. Seed rate is 1-1.5 kg/ha. Seed can be sown in situ or in a nursery seedbed and then transplanted to their permanent position 2 - 3 weeks later. Germination is usually rapid and good if the soil is warm, but poor germination rates are experienced in cool or cold soils. A drop in temperature overnight aids in germination. Cuttings of growing plants root easily.

Plants should not be given inorganic fertilizers. Biofertilizer requirement is high. For good yield its need inorganic fertilizer. Application of neemcake is beneficial to protect from insect and pest. Timely weeding and hoeing is required for this plant. Irrigation and fertilizer application is needed.

14. Common name: Telakuchu sak, Ivy Gourd, Scarlet Gourd, Tindora

Scientific name: *Coccinia cordifolia*

Family: Cucurbitaceae

INTRODUCTION

The plant is sporadically cultivated for its edible fruit and also its medicinal uses in tropical and sub-tropical zones, particularly in India, Malaysia and Indonesia. The plant is sold in local markets, and is also occasionally exported to speciality markets. There are some named varieties.

HABITAT

Habitats are deciduous bush, savannah, dry evergreen forest, thickets, moist neglected places, especially on hedges.

MEDICINAL PROPERTIES

Its can protect us from allergy, jaundice, diabetic development, etc. It is used as a blood purifier and acts as an insect repellent. The root extract of Coccinia also protects from breathing problem and fever. The juice of the stem is dripped into the eyes to treat cataracts. The fruit has sometimes been used as an eye-medicine. An infusion of the leaves is given to women to drink when a birth is overdue. The leaves are used externally as a poultice in treating skin eruptions. The plant is laxative. It is used internally in the treatment of gonorrhoea. A decoction of the plant is used to treat chest colds. The plant is crushed and applied externally to affected areas in order to alleviate the pain of headache and rheumatism.

ANTI-DIABETIC ACTIVITY

Hypoglycemic effect has been observed in patients treated with homogenized freeze dried leaves of *Coccinia indica.* Ethanol extract (250mg/kg) of whole plant of *Coccinia indica* reduce the blood glucose level in fasted, glucose fed and diabetic albino rats. Alcoholic leaf extract of *Coccinia indica* produced hypoglycemic effect in normal fed and 48 hours fasted rats and response was mediated by suppression of gluconeogenic enzyme glucose-6-phosphatase.

Antimicrobial Activity

The aqueous extracts of seeds of *Coccinia indica* did not show much significant activity, while the organic extracts (petroleum ether and methanol) showed the highest activity against *staphylococcus aureus*. *In vitro* antibacterial activity of leaves and stem extracts of *Coccinia grandis* L. has been investigated against *Corynebacterium diptheriae, Bacillus cereus, Streptococcus pyogenes, Staphylococcus aureus, Escherichia coli, Proteus mirabilis, Klebsiella pneumonia, Salmonella typhi, Pseudomonas aeruginosa* and *Shigella boydii*.

Anti-Inflammatory Activity

Both post- and pre-treatment anti-inflammatory activities of the aqueous extract of fresh leaves of *Coccinia indica* were evaluated in rats at various dose levels using the carrageenan-induced paw oedema method. In post-treatment studies, a dose-dependent anti-inflammatory effect was observed in the dose range of 25–300 mg/kg which was equivalent to diclofenac (20 mg/kg) at 50 mg/kg.

Analgesic and Antipyretic Activity

Analgesic and antipyretic properties were evaluated using tail flick model and yeast-induced hyperpyrexia, respectively. The extract produced noticeable analgesic activity as good as to morphine at 300 mg/kg, which suggests the connection of central mechanisms. Methanolic extract of *C. cordifolia* for antipyretic activity at the doses of 100 and 200 mg/kg in yeast-induced fever showed antipyretic activity by elevating the prostaglandin (a regulator of body temperature) biosynthesis.

Antioxidant Activity

The antioxidant activities of the various fractions of the hydromethanolic extract of the leaves of *Coccinia grandis* L. have been well documented. The roots, stems, leaves and whole plant of *C. grandis* are used in the treatment of jaundice, bronchitis, skin eruptions, burns, insect bites, fever, indigestion, nausea, eye infections, allergy, syphilis, gonorrhoea, etc.

Other Uses

Young leaves and long slender stem tops are cooked and eaten as a potherb or added to soups. Young and tender green fruits are consumed as raw in salads or cooked and added to curries etc. Ripe scarlet fruits are eaten as raw. The pulp of fully ripe fruits is fleshy, creamy and with an agreeable slightly acid flavour. The smooth and thin-skinned egg shaped fruit turns red when fully ripe. Another species is also important namely Kenduri. It is mostly seen in North and South India, looks like a pointed gourd and good for health.

Botany

Ivy gourd is a perennial climbing plant producing stems up to 20 metres long from a tuberous rootstock. It clambers over other plants, supporting itself by means of tendrils. Young stems are annual and herbaceous, but can become perennial and woody with time. This plant is a perennial climber with single tendrils and glabrous leaves. The leaves have 5 lobes and are 6.5–8.5 cm long and 7–8 cm wide. The species is dioecious. Female and male flowers emerge at the axils on the petiole, and have 3 stamens. The fruits can be up to 5cm long.

Chemical Constituents

The fruits of *C. cordifolia* contain active constituents like taraxerone, taxerol, amyran, lupeol and glycoside cucurbitacin B. The leaves are bitter sweet and astringent. Major phytoconstituents present in *C. cordifolia* are cardenolides, saponins, flavonoids and polyphenols. Seed fat mainly contains palmitic (16.3%), oleic (22.4%) and linoleic (58.6%) acids. Plant also contains arabinogalactan, xyloglucan and xylan. Phytochemical analysis showed that petroleum ether extract contains sterol compound and tannins, proteins and amino acid as well as glycosides were present in chloroform extract.

Agrotechniques

The plant has escaped from cultivation and become naturalized in several areas. A noxious vine, it smothers vegetation and other objects forming a dense canopy. It acts as a host for melon fly and is a reservoir for other crop pests possibly including ring spot virus. It can very well succeed in most soils but prefers a sunny sheltered position in a humus-rich open sandy soil. It very much dislikes clay soils. The plants are to be kept well watered throughout the growing season. A dioecious plant, both male and female forms must be grown if seed and fruits are required.

15. Common name: Thankuni, Indian Pennywort, Centella, Gotu Kola

Scientific name: *Centella asiatica*

Family: Apiaceae

Introduction

It is an important medicinal herb of the Orient. With a history of use since thousands of years, it is a relished vegetable and drink in parts of Southeast Asia. It is also becoming more and more admired in other areas of the world, particularly for its curative merits. The plant is normally gathered from the wild, but is also sometimes cultivated both for its medicinal uses and its edible leaves. It is frequently traded, both locally and internationally, and is a foremost component of many over-the-counter medicinal preparations.

Habitat

Centella asiatica usually grows in fresh water wetlands, wet ditches and margins of ponds.

Medicinal Properties

It acts as a blood purifier and protects from vomiting trouble. It is anti allergic. The compounds brahmoside and brahminoside demonstrate antispasmodic, antipyretic, central nervous system-depressant and hypotensive activity. It is also having promising narcotic analgesic activity mediated through opiodergic receptors. The ethanolic extract has revealed anti-stress activity as good as to that of diazepam. Various trials with patients suffering from venous insufficiency of the lower limbs have shown a significant difference in favour of a titrated extract of the plant for the symptoms of heaviness in the lower limbs and oedema. An aqueous extract of the plant showed activity against the herpes simplex II virus. The compound encourages healing through bacteriostatic activity and stimulates of the reticuloendothelium. To cure wounds, haemorrhoids and rheumatic joints, the herb is often applied externally. The crushed leaves or extracts of the plant are applied topically in the add-on treatment of surgical wounds and minor burns. The extract is used effectively in the treatment of keloids, leg ulcers, phlebitis, slow-healing wounds, scleroderma, lupus, leprosy, surgical lesions, striaedistensae, cellulitis and aphthae. Purified extracts are known to speed up cicatrizing and skin grafting.

Other Uses

Leaves having slightly bitter flavour are consumed raw or cooked as a vegetable. They are used in salads, cooked in curries, soups and stews. By diluting the leaf juice with water and adding sugar, a stimulating summer drink can be prepared. A tonic tea is also made from the leaves.

Botany

Gotu kola is a low-growing, evergreen perennial prostrate herb, rooting at the nodes and about 20cm tall. The plant spreads quickly by the roots, producing long stolons up to 250 cm in length. These roots at the nodes and the plant can soon form large carpets of growth. Leaves are usually glabrous, orbicular-reniform, entire, crenate or lobulate. Bracts are small, ovate. Flowers are white, borne in 3-6 flowered umbels. Fruits and seeds are laterally compressed. Flowering and fruiting occurs from April-June.

Chemical Constituents

Presence of several flavonoid derivatives such as quercetin, kaempferol, patuletin, rutin, apigenin, castilliferol, castillicetin, and myricetin has been reported in *C. asiatica*, while isolation of polysaccharides (e.g., centellose), polyacetylenes (e.g., cadinol, acetoxycentellinol, centellin, centellicin, and asiaticin), sterols (e.g., 11-oxoheneicosanil-cyclohexane, dotriacont-8-en-1-oic acid, sitosterol 3-O-β-glucoside, stigmasterol 3-O-β-glucoside, and castasterone), and phenolic acids (e.g., rosmarinic acid, 3,5-di- O-caffeoyl quinic acid, 1,5-di- O-caffeoyl quinic acid, 3,4-di- O-caffeoyl quinic acid, 4,5-di- O-caffeoyl quinic acid, ettacrynic acid, chlorogenic acid, and isochlorogenic acid has been also identified. Several other phenolic acids, for example, *p*-hydroxybenzoic acid, vanillic acid, *p*-coumaric acid, *o*-coumaric acid, and *trans*-cinnamic acid are also found.

Agrotechniques

It prefers a moist to wet soil in sun or partial shade. The plant can grow and flower all the year round in moist climates, though it dies down in the dry season in monsoon climates. Division can very easily be done to propagate plants at any time in the growing season. Stem pieces with one node are planted directly in the field or first in a nursery. The seedlings are transplanted into individual pots when they are large enough to handle and grown on until large enough to plant out. Shading is not necessary, but sufficient soil moisture is essential. If conditions are positive, the first harvest can be obtained 2-3 months after planting. Succeeding harvests are likely in every two months if only leaves are to be harvested. However, creeping types are usually harvested as whole plants. For medicinal use, whole plants are generally harvested at any suitable time of the year. The yield of bushy forms may reach 8 tonnes per hectare of fresh leaves for the first harvest, and 14 tonnes for each subsequent harvest. Yields of the creeping types are lower. Fresh leaves harvested as a vegetable are tied together in small bundles and need to be marketed quickly, as they wilt rapidly.

16. Common name: Dhone, Coriander, Cilantro, Chinese parsley
Scientific name: *Coriandrum sativum*
Family: Apiaceae

INTRODUCTION

Coriander is basically a seed spice. But the leaves are used as leafy vegetables especially to add spicy flavour and attractive taste in our day to day cooking. Both the seeds and the herb of coriander are used as spice or a medicinal plant. It is popular throughout the world.

HABITAT

Coriander, an umbelliferous plant native to southern Europe, is found infrequently in fields and waste places, and by the sides of rivers and anthropogenic (man-made or disturbed habitats) meadows. It is frequently found in a semi-wild state in the east of England, having escaped from cultivation. It can grow in semi-shade (light woodland) or no shade. It prefers dry or moist soil.

MEDICINAL PROPERTIES

Antibacterial and Antifungal Properties

The effective antibacterial activity of *C. sativum* essential oil against *Staphylococcus aureus* and Gram-negative bacterial strains including *Escherichia coli*, *Klebsiella pneumoniae*, *Salmonella typhimurium*, and *Pseudomonas aeruginosa* and two clinical multidrug-resistant *Acinetobacter baumannii* isolates has been shown. Aliphatic (2E)-alkenals and alkanals characterized from the fresh leaves of *C. sativum* were found to possess bactericidal activity against the food-borne bacterium, *Salmonella choleraesuis* subsp. *choleraesuis*, respectively.

Antioxidant Activity

There are ample evidences that coriander has efficient anti-oxidative characteristics, consequently linked to anti-cancer, hypolipidemic, anti aging, and anti-inflammatory activities.

Antidiabetic Effects

Sub-chronic oral administration of *C. sativum* extract (20 mg/kg) in obese-hyperglycemic and hyperlipidemic animal model normalized glycemia and decreased the elevated levels of insulin, insulin resistance (IR), total cholesterol (TC), low density lipoprotein (LDL)-cholesterol, and triglycerides (TG).

Cardioprotective Effects

Since *C. sativum* extract decreased several components of the metabolic syndrome and decreased atherosclerotic and increased cardioprotective indices, its extract may have cardiovascular protective effect. It has been demonstrated that *C. sativum* extract was able to decrease hyperglycemia and increase glucose uptake and metabolism, and insulin secretion.

OTHER USES

Leaves are used to impart aromatic flavouring in salads, soups etc. It is foetid according to another report. The fresh leaves have also reported to be with a strong bedbug-like smell. The fresh seed has a disagreeable and nauseous smell, but when dried it becomes fragrant, the longer it is kept the more fragrant it becomes. Seeds are cooked and used as flavouring in many dishes including cakes, bread and curries. They are also widely used to flavour certain alcoholic liquors. The root is powdered and used as a condiment. An essential oil from the seed is used as food flavouring.

Botany

Coriander (*Coriandrum sativum* L.) is a plant from the family of Apiaceae. It is an annual growing to 0.5 m by 0.3 m. The species is hermaphrodite (has both male and female organs) and is pollinated by insects. The plant is self-fertile.

Chemical Constituents

The fresh leaves contain about 0.012% oxalic acid and 0.172% calcium. Aliphatic aldehydes (mainly C10 to C16 aldehydes) with fetid-like aroma are predominant in the fresh herb oil. Major components in the oil isolated from coriander fruit include oxygenated monoterpenes and monoterpene hydrocarbons. The essential oil content of dried coriander fruits varies between 0.03 and 2.6%, while the fatty oil content varies between 9.9 and 27.7%. The essential oil contains flavoring compounds such as linalool, geraniol, pinene, limonene, geranyl acetate, terpinene, and borneol. Other constituents including crude protein, fat, crude fiber, and ash contents vary from 11.5 to 21.3%, 17.8 to 19.15%, 28.4 to 29.1%, and 4.9 to 6.0%, respectively.

Agrotechniques

It is suitable for light (sandy) and medium (loamy) soils and prefers well-drained soil. Suitable pH is acidic, neutral and basic (alkaline) soils and can grow in very alkaline soils. Good quality seed is needed to cultivate coriander. For leaf production it can be cultivated throughout the year. The field should be friable. Seed is schizocarp and should be broken into two halves. Before swing, seeds are put in cold water for a minimum of 12 hours. After sowing of seed, dry cow dung can be applied for better growth of seed with required amount of irrigation. After 7-10 days, germination starts and after one month thinning of the seedlings can be done. Application of irrigation is need-based. Coriander leaf cultivation is profitable as a market demand.

17. Common name: Nune, Parsley, Duckweed, Little hogweed
Scientific name: *Portulaca oleracea*
Family: Portulacaceae

INTRODUCTION

It is perennial and soft in character. Flowering occurs in rainy season and fruits mature in winter. It is a prostrate, mat-forming annual or short-lived perennial herb with much-branched, spreading, articulated, fleshy stems up to 30 cm long or longer. The stems root at the nodes where they are in contact with the ground. The edible leaves are gathered from the wild and used locally. The plant is also used for local medicinal purposes.

Habitat

Nunesak can grow in cannel, pond side area and also in fallow land. It prefers open disturbed grounds on sandy soils.

Medicinal properties

- It is diuretic, sedative, analgesic, tonic, choleretic and cardiotonic,
- It can treat rheumatism and gynaecological diseases, fever, disorders of the urinary tract, worm diseases, dysentery,
- For ulcers, eczema and dermatitis, it is applied externally.
- It protects from piles, skin disease, arthritis and bile problems.
- Two spoon juice of portulaca leaf is effective against stammering.
- One drop juice of portulaca can be used directly in eye for curing conjunctivitis.
- 10-15 drops of portulaca mixed with honey protect from cold and cough of children.
- 5-15 drops of portulaca can protect from dysentery.

Other Uses

The leaves and young shoots are collected from the wild and are eaten raw. They have a gentle agreeable flavour and are regularly used in salads. They are also consumed as a cooked vegetable. In India, boiled leaves are mixed with sorghum or pearl millet flour in preparing a kind of bread.

Botany

Prostrate, mat-forming annual or short-lived perennial herb with much-branched, spreading, articulated, fleshy stems up to 30 cm long or longer, rooting freely from the nodes, often flushed reddish; nodes with a dense whorl of whitish hairs. Leaves opposite, simple, sessile, narrowly elliptical to ovate, 0.5–1.5 cm × 1–4 mm, apex obtuse to subacute, smooth, veins distinct. Flowers solitary at the tips of short lateral branches, surrounded by 4 involucral leaves and copious hairs, bisexual; sepals 2, ovate, 3(–6) mm long; petals 4, obovate, 3.5–10 mm × 4 mm, usually yellow (rarely pinkish); stamens 7–16, arranged in 1 whorl; ovary half-inferior, style usually with 4 arms. Fruit an obovoid capsule 2–3.5 mm long, dehiscing near the base leaving only a very thin persistent rim, many-seeded. Seeds are semi-orbicular in outline,1 mm in diameter, dull grey.

Chemical Constituents

Many constituents of *Portulaca oleracea* have been isolated, including flavonoids, alkaloids, fatty acids, terpenoids, polysaccharides, vitamins, sterols, proteins, and minerals. In addition to flavonoids, another important chemical found in this plant is alkaloids including dopa, dopamine, and noradrenalin. The content of dopamine and noradrenalin is higher in leaves compared to stem and seeds.

Agrotechniques

The plant is widespread in the tropics. It grows best in areas where annual daytime temperatures are within the range 22 - 32°C. It prefers a mean annual rainfall in the range 1,500 - 2,000 mm. It requires a sunny position in a well-drained sandy or sandy loams soil. Plants can succeed in quite poor soils. Prefers a pH in the range 5.5 - 6.5, tolerating 4.5 - 7.5.Plants can become troublesome weeds of cultivation, since they break up easily and even small fragments can grow into new plants. In addition, the seeds are easily spread by wind, water, with crop seeds or through bird droppings. The plants take about six to eight weeks to produce a crop from seed and can then be harvested every two weeks on a cut and come again principle. The flowers are said to open promptly at 10am, hence the English name 'ten o'clock plant'.

18. Common name: Punarnaba, Red spiderling, Spreading hogweed
Scientific name: *Boerhaavia diffusa*
Family: Nyctaginaceae

INTRODUCTION

Boerhaavia diffusa Linn. is a well-known medicinal plant in traditional Indian medicine as well as other parts of world, viz., Southern American and African continents. Its various parts and especially roots have been used for gastro-intestinal, hepatoprotective, and gynecological indications in above mentioned parts of the world and also throughout India. In ayurvedic texts, more than 35 formulations of different types contain it as one of the key ingredients.

It is a perennial shrub with huge branching habit rich in vitamins and protein. The root is hard and stem is trailing. It is a very popular medicinal plant, especially in India, where it is widely used in Ayurveda. Medicines containing this plant are sold worldwide. The plant is gathered from the wild and is also sometimes used as food.

Habitat

It is a basically a weed found abundantly throughout India. It is a perennial, spreading hogweed, commonly occurring abundantly in waste places, ditches and marshy places during rains. This leafy vegetable is found in barren land, beside railway tracks, open places near the sea, dry and warm river valleys, along roadsides, near habitations, in and along cultivated fields and in open cleared patches in forests, etc.

Medicinal Properties

Anxiolytic Activity

Anxiolytic activity of hydroalcoholic extract of *B. diffusa* leaves was evaluated in rats. The results revealed that hydroalcoholic extracts of *B. diffusa* had significant anxiolytic activity which was comparable to standard drug Diazepam (0.5 mg/kg).

Protective Effect on Gastrointestinal Problems

The anti-ulcer activities of *B. diffusa* are contributed by several mechanisms such as its healing effects, regulation of gastric secretions, anti-inflammatory properties and mucus stimulatory effects.

Anticancer Activity

Radiotherapy, surgery and chemotherapy methods used to treat cancer are highly effective but exert numerous side effects. Also the cancer cells gradually become resistant to the treatments. To avoid it, scientists are doing numerous efforts to find herbal solutions to cancer. Anticancer activity of ethanol extract of *B. diffusa* leaves against Dalton's ascitic lymphoma in mice is reported.

Anti-Diabetic Property

Indigenous system of medicine in India has recommended many herbal medicines including *B. diffusa* for curing diabetes. It was found that there was some decrease in 24 hour urine protein in the patients after six months follow up.

Other Uses

Long, thick and fleshy roots are eaten by the Aborigines of central Australia. Leaves are cooked as a pot herb or added to soups. It is also occasionally used as vegetable. Seeds are cooked. Greenish white plant is used in piles, cold/cough, gas problem, indigestion. Bluish plant is used in breathing problem, beri beri, heart desease, etc. Stem juice is used to protect from jaundice, urine problem, ganoria etc.

Botany

Boerhaavia diffusa is a herbaceous perennial plant with vigorous, many-branched stems growing from a taproot; up to 2 metres long the stems can be erect or procumbent. The stems branch mainly from the base, they are prostrate when young, becoming ascending to erect when flowering. Leaves are unequal, ovate, blunt, wavy along margins, flat to somewhat heart-shaped at base, woolly; leaf-stalk is 1 cm long. Inflorescences occur at the end of branches, are forked about 3-6 times, occasionally with sticky internodal bands. Branches are divergent, terminating in compact subumbellate or capitate, 2-5-flowered clusters.

Flowers have stalk shorter than 0.5 mm. Bract at base of the flower tube quickly deciduous, lancelike, 0.8-1 mm. Flowers are purplish red to reddish pink or nearly white, bell-shaped beyond the constriction, 1-1.5 mm. Stamens 2-3, are inside the flower or barely protruding out. Tender young leaves and shoots are cooked and used as a vegetable. Red Spiderling is a plant found throughout India.

Chemical Constituents

Generally whole plant consists of phytochemicals like punarnavine (alkaloids), ß -sitosterol (phytosterols), liriodendrin (lignans), punarnavoside (rotenoids), boerhavine (xanthones) and potassium nitrate (salts). The roots contain the rotenoids boeravinones AI, BI, C2, D, E and F besides the new dihydro-isofurenoxanthin, alanine, arachidic acid, aspartic acid, behenic acid, boerhaavic acid, boerhavone, campesterol, daucosterol, beta-ecdysone, flavone, 5-7-dihydroxy-3'-4'-dimetho, XY-6-8-dimethyl, galactose, glutamic acid, glutamine, glycerol, glycine, hentriacontane N,heptadecyclic acid, histidine, hypoxanthine-9-l-arabinofuranoside, leucine, liriodendrin, methionine, oleic acid, oxalic acid, palmitic acid, proline, hydroxyl serine, sitosterol oleate, sitosterol palmitate.

Agrotechniques

The plant is a weed of cultivated land and wasteland, often in lawns in drier areas. It prefers a sunny position and a well-drained soil. The plant grows wild in a range of soil types. Soil with good moisture and-semi shady condition is better for its cultivation. Medium or high land is better for its cultivation. March-April month is ideal for its cultivation. Much more care is not needed for its cultivation. Application of Urea followed by one or two weeding is better for its proper growth and development.

Boerhavia diffusa is widespread through much of the tropics and the subtropics and has also become naturalized in parts of the temperate zone. Plants can commence flowering when only four weeks old from seed. The plant can flower and produce fruit all the year round if sufficient water is available.

19. Common name: Palta, Pointed gourd, Parwal
Scientific name: *Trichosanthes dioica*
Family: Cucurbitaceae

INTRODUCTION

Pointed gourd is a perennial, climbing plant producing stems to sprawl over the ground or climb into other plants for support, attaching themselves by means of tendrils. Its edible fruit is highly prized in India. The species is well-known in Indian system of medicine. The plant has been cultivated, especially in eastern India, for over 2,000 years. It is also cultivated in Japan, Sri Lanka, China, and Thailand.

Habitat

Pointed gourd is a tropical vegetable crop. It is believed to be originated in the Indo-Malaya region. The crop is distributed extensively in eastern parts of India and to a lesser extent, in rest parts of Asia.

Medicinal Properties

Antidiabetic Activity

The aqueous extract of *T. dioica* leaves has good hypoglycaemic potential along with a high antidiabetic profile. In rats with streptozotocin-induced severe diabetes mellitus, aqueous extract of *T. dioica* fruits dose of 1000 mg/kg body weight daily once for 28 days reduced the levels of fasting blood glucose, postprandial glucose, aspirate aminotransferase, alanine aminotransferase, alkaline phosphatase, creatinine, urine sugar and urine protein whereas total protein and body weight was increased. No toxic effect was observed during LD50.

Gastric Antiulcer Activity

The Aqueous extract of leaves of *Trichosanthes dioica* was evaluated for its antiulcer activity in Wistar rats. Ranitidine (100 mg/kg) was used as the standard drug. Different groups of rats were given two doses (250 and 500 mg/kg) of *T. dioica* extract. Phytochemical results revealed the presence of tannins, saponins, triterpenoids, flavonoids which confirmed antiulcer activity.

Cholesterol-Lowering Activity

The effect of *T. dioica* fruit extracts on cholesterol and triglycerides were studied in rats. *T. dioica* treatment caused a significant decrease in plasma cholesterol levels after a single administration and after repeated oral administrations.

Anti-Oxidant Activity

Anti-oxidant activity of aqueous extract of *Trichosanthes dioica* (TSD) fruits was studied for its free radical scavenging property. The findings could justify the inclusion of this plant in the management of antioxidant activity.

OTHER USES

The fruits, being an integral part of an average Indian diet, are consumed as a vegetable. Hence, it would not be wrong to state that still a lot has to be worked upon this important plant. Apart from this, old traditional literature like Charak Samhita mentioned the protective role of *Trichosanthes dioica* on important body organs like liver, spleen, *etc.*, many of which are now scientifically proved.

BOTANY

The plant is a perennial, dioecious, and grows as a vine. Roots are tuberous with long taproot system. Vines are pencil thick in size with dark green cordate simple leaves. Flowers are tubular white with 16–19 days initiation to anthesis time for pistillate flowers and 10–14 days for staminate flowers. Stigma remains viable for approximately 14 hours and

40–70% of flowers set fruit. Based on shape, size and striation, fruits can be grouped into 4 categories: (1) long, dark green with white stripes, 10–13 cm long, (2) thick, dark green with very pale green stripes, 10–16 cm long, (3) roundish, dark green with white stripe, 5–8 cm long, and (4) tapering, green and striped, 5–8 cm long6.

Chemical Constituents

T. dioica is rich in vitamins (A and C), tannins, saponins and contains 9.0 mg Mg, 2.6 mg Na, 83.0 mg K, 1.1 mg Cu, and 17.0 mg S per 100 g edible part. Two main phytosterols present in *T. dioica* are namely, 2-4-α-ethylcholest-7-enol and 2-4-β-ethylcholest-7-enol. Seeds of *T. dioica* contain lectin, a carbohydrate (specifically galactose) binding protein which is homologous to Type-II ribosome inhibitory proteins (Type-II RIP).

Agrotechniques

T. dioica is usually propagated through vine cuttings and root suckers. Seeds are not used in planting because of poor germination and inability to determine the sex of plants before flowering. As a result, crop established from seed may contain 50% nonfruiting male plants. Both pre-rooted and fresh vine cuttings are used for propagation. Fresh vines used for field planting should have 8–10 nodes per cutting at a spacing of 1.5–2.0 m × 1.5–2.0 m. A female: male ratio of 9:1 is most favourable for ensuring maximum fruit set. FYM and rotten leaf can be used as a fertilizer at the time of field preparation. After sowing of seeds, the land is covered with coconut leaf followed by a light irrigation. Careful harvesting is done after few days and marketed.

20. Common name: Pat, Tossa jute, Jew's mallow, West African sorrel

Scientific name: *Corchorus olitorius*

Family: Malvaceae

INTRODUCTION

The plant is a very popular vegetable crop in many tropical countries, where it makes an excellent substitute of spinach in areas with hot summers. It is often cultivated in the tropics and warm temperate areas both for the fibre in its stem and also for its edible leaves.

HABITAT

The plant grows well in the lowland tropics, ranging from warm temperate zones through tropical desert to wet forest life zones.

Medicinal Properties

Tender leaf is tasty enough to be used in culinary purpose. It protects from cold/cough, arthritis, acidity, skin disease and dysentery. The leaves are also demulcent, diuretic, febrifuge and tonic. They are used in the treatment of chronic cystitis, gonorrhoea and dysuria.A cold infusion is said to restore the appetite and strength. The seeds are purgative. Injections of olitoriside, an extract from the plant, markedly improve cardiac insufficiencies and have no cumulative attributes; hence, it can serve as a substitute for strophanthin.

Other Uses

The fibre of jute, obtained from the stems, is somewhat coarse and is used mainly for sackcloth etc. The stems are harvested when the plant is in flower and are then retted to extract fibre. If used in making paper, the fibres are cooked for 2 hours with lye and then ball milled for 4½ hours. The paper is grey/buff. The very light and soft wood is used in making sulphur matches.

Botany

Corchorus olitorius is an erect herbaceous annual to short-lived perennial plant, fairly branched and grows about 1.5 m high. The plant carries the flowers solitary or in two-flowered cymes opposite of the leaf. The flowers, borne on the end of a short stem, count 5 sepals, 5 petals and 10 free and yellow stamina.

CHEMICAL CONSTITUENTS

The leaves are reported to contain 43-58 calories, 80.4-84.1 g water, 4.5-5.6 g protein, 0.3 g fat, 7.6-12.4 g total carbohydrate, 1.7-2.0 g fiber, 2.4 g ash, 266-366 mg Ca, 97-122 mg P, 7.2-7.7 mg Fe, 12 mg Na, 444 mg K, 6,410-7,850 μg beta-carotene equivalent, 0.13-0.15 mg thiamine, 0.26-0.53 mg riboflavin, 1.1-1.2 mg niacin, and 53-80 mg ascorbic acid per 100 g. Leaves contain oxidase and chlorogenic acid. The folic acid content is substantially higher than that of other folacin-rich vegetables, 800 micrograms per 100 g (75% moisture) or 3200 micrograms on a zero moisture basis. The seeds contain 11.3-14.8% oil, reportedly estrogenic, which contains 16.9% palmitic-, 3.7% stearic-, 1.8% behenic-, 1.1% lignoceiic-, 9.1% oleic-, 62.5% linoleic-, and 0.9% linolenic- acids as well as large portions of B, Mn, Mo, and Zn.

AGROTECHNIQUES

Plants grow well in the lowland tropics, up to an elevation of around 700 metres. It can tolerate an annual precipitation between 400 and 4290 mm, an annual average temperature range of 16.8 to 27.5°C. It prefers a very fertile, humus-rich, well-drained alluvial soil having a pH in the range of 4.5 to 8.2. It grows best in a hot humid climate. It does not tolerate extremely waterlogged soils and some cultivars are sensitive to excess water in the soil, especially when they are young.

This species tends to branch thereby making fibre extraction more difficult. Growing the plants very close together will prevent some of the branching. The first harvest, by cutting shoots 20 - 30cm long, may take place 4 - 6 weeks after transplanting, at a height of 10 - 20cm above the ground. This cutting stimulates the development of side shoots. Subsequently, every 2 - 3 weeks, a cutting may take place, with a total of 2 - 8 cuttings possible. For an once-over harvest from a direct sown crop, the plants are uprooted or cut at ground level when they are 30 - 40cm tall, 3 - 5 weeks after emergence and before the development of fruits.

21. Common name: Pui, Malabar spinach, Ceylon spinach, Vine spinach

Scientific name: *Basella alba*

Family: Basellaceae

INTRODUCTION

Basella alba is a short-lived, perennial climbing plant, producing stems up to 9 metres long that scramble over the ground and twine into other plants for support. The plant is widely cultivated for its edible leaves in the tropics. These can be harvested all the year round, there are some named varieties. The leaves are sometimes sold in local markets. Red forms of the plant are often grown as ornamentals.

HABITAT

Moist places in hedges at elevations around 500 metres.

Medicinal Properties

Basella or *Pui sak* is a very succulent vegetable. It protects us from cough and cold. The roots are astringent. They are cooked and used in the treatment of diarrhoea. A paste of the root is applied to swellings and is also used as a rubefacient. The leaves and stems are cooked and eaten for their laxative properties. The leaf juice is a demulcent, used in cases of dysentery. It is also diuretic, febrifuge and laxative. The leaf juice is used in Nepal to treat catarrh. The red juice of the fruit is used as eye drops to treat conjunctivitis. Leaf paste is applied externally to treat boils and sores. The flowers are used as an antidote to poisons. Its juice is a safe aperient for pregnant women and a decoction has been used to alleviate labour pain.

Other Uses

Leaves and stem tips are consumed raw or cooked. The mucilaginous qualities of the plant make it an excellent thickening agent in soups, stews etc., where it can be used as a substitute for okra. The leaves are a good source of vitamins A and C, iron and calcium. An infusion of the leaves is a tea substitute. The purplish sap from the fruit is used as a food colouring in pastries and sweets. The colour is enhanced by adding some lemon juice. It has been used as rouge and also as a dye for official seals.

Botany

Glabrous annual or shortly lived perennial, succulent tangled twiner; stems much branched, 2-10 m long, sometimes almost leafless, greenish or reddish. Leaf-lamina ovate is to suborbicular, 5-15 cm long, 5-13.5 cm broad, acute or acuminate (less commonly obtuse), usually widely cordate at the base; lateral nerves 4-5 on either side; petiole 2.5-6.5 cm long. Flowers white, rose or purplish, 4-5 mm long, in long-peduncled spikes,

2.5-15 cm long, usually unbranched but branched in some cultivated forms. Perianth is fleshy, urceolate, somewhat saccate at the base; lobes short, ovate, about one-third the length of the tube, not opening.Fruits0.5 cm in diameter (4-7 x 5-10) mm, red, white or black; surface crinkly in the dry state.

Chemical Constituents

Dried leaves contains, protein 20%, fat 3.5%, carbohydrate 54%, fibre 9%, and ash 19% per 100 gm. The leaves contain a high level of calcium and are rich in vitamins like A, C, thaimine, riboflavin, niacin, and betacyanin, oxalic acid, flavonoid like acacetin, 4,7- dihydroxy kempferol and 4'-methoxyisovitexin and also phenolic acids like vanilla, syringic and ferulic acid. The fruit contains betacyanins and gomphrenin.

Agrotechniques

Basella alba is a plant of the moist tropics, where it is found at elevations up to 1,500 metres. It grows best in areas where annual daytime temperatures are within the range 23 - 27°C. It is not tolerant of frost. It prefers a mean annual rainfall in the range 2,000 - 2,500 mm. Best growth is produced in a well-drained moisture-retentive soil rich in organic matter with a pH in the range 5.5 – 7 and a warm sunny sheltered position. Plants produce larger, more succulent leaves when grown in light shade. As a fast growing insect and disease resistant plant, it is capable of producing a crop within 70 days from seed in a warm climate.

The plant can be grown on living stakes, usually on a fence or on a hedge. Plants do not flower if the length of daylight is more than 13 hours per day. A vigorous, fast-growing plant, when established it can be harvested every week or so by removing young shoots 7 – 13 cm long. This encourages the production of side shoots and thus more stems to harvest. A short-term crop yields upto 40 tonnes per hectare in 75 days. A long-term crop, yields upto 1.5 kg of shoots or leaves per plant or 80 tonnes in 180 days.

22. Common name: Palong, Common spinach, Spinach
Scientific name: *Spinacia oleracea*
Family: Amaranthaceae

INTRODUCTION

Spinacia oleracea is an annual plant. It contains leaf throughout the year, but flowers from June to September. The species is dioecious (individual flowers are either male or female, but only one sex is to be found on any one plant so both male and female plants must be grown if seed is required) and is pollinated by wind. The plant is not self-fertile.

HABITAT

Spinacia oleracea (Spinach) is a species of annual herb in the family pigweed. It has a self-supporting growth form. It is associated with freshwater habitat. Spinach has simple, broad leaves.

MEDICINAL PROPERTIES

Spinach is considered as a heart-healthy food as it is loaded with vitamins C, A and folate which are antioxidants. The leaves are cooling,emollient,wholesome, antipyretic, diuretic,laxative, anthelmentic, useful in urinary concretion, inflammation of the lungs and the bowels, sorethroat,pain in joints, thirst, lumbago,cold and sneezing,soreeye, ringworm,scabies, leucoderma, biliousness, flatulence, etc. It can check blood pressure and cardiovascular disease. The magnesium contained in spinach helps prevent diabetes. Spinach has excellent anti-inflammatory properties and can reduce migraine, headache as well as asthma. Fresh spinach abounds in cancer-fighting antioxidants like flavonoids and kaempferol. Thecarotenoid found in spinach that is helpful against cataract and macular degeneration in eye. Crushed spinach leaves are reported to prevent kidney or bladder diseases. Seeds have been used in the treatment of difficulty in breathing and jaundice.

BOTANY

Spinacia oleracea is an annual plant growing to a height of 50 centimeters. Leaves are succulent, alternate, simple, ovate or triangular-based, varying in sizes from 2 to 30 centimeters long and up to 15 centimeters wide, larger at the base of the plant, smaller on flowering stems. Flowers are inconspicuous, yellowish green, 3 to 4 millimiters in diameter, maturing to a small, hard, dry and lumpy fruit cluster about 5-10 millimeters across, containing several seeds. There are two main types of spinach leaves:

- the crinkled savory-leaf spinach and
- the smooth or flat-leaf spinach.

Chemical Constituents

Nutrient analysis of raw spinach per 100 g: (Proximates) water 91.40 g, energy 23 kcal, protein 2.86 g, total lipid (fat) 0.39 g, carbohydrate by difference 3.63 g, total dietary fiber 2.2 g, total sugars 0.42 g; (Minerals) calcium 99 mg, iron 2.71 mg, magnesium 79 mg, phosphorus 49 mg, potassium 558 mg, sodium 79 mg, zinc 0.53 mg; (Vitamins) vitamin C (total ascorbic acid) 28.1 mg, thiamin 0.078 mg, riboflavin 0.189 mg, niacin 0.724 mg, vitamin B6 0.195 mg, folate 194 μg, vitamin A 9377 IU, vitamin E 2.03 mg, vitamin K 482.9 μg; (Lipids) total saturated fatty acids 0.063 g, total monosaturated FA 0.010 g, total polyunsaturated FA 0.165 g. Phytochemical screening yielded phlobatannin, saponins, phenol, tannins, glycosides, flavonoids, steroids, terpenes, and cardenolides. Important compounds are: phenolics (ortho-coumaric, and ferulic acid); flavonoids (apigenin, glucuronide, flavone, jaceidin, kaempferol, myrcetin, methoxyflavone, patuletin, quercetin and spinacetrin); carotenoids (lutein, β-carotene, violaxanthin, and neoxanthin), vitamins and minerals, etc.

Agrotechniques

Propagation: Seed - sow in situ from March to June for a summer crop. Make successional sowings, perhaps once a month, to ensure a continuity of supply. The seed germinates within about 2 weeks and the first leaves can be harvested about 6 weeks later. Seed is sown in situ during August and September for a winter crop.

23. Common name: Pudina, Spearmint, Garden mint
Scientific name: *Mentha spicata*
Family: Lamiaceae

INTRODUCTION

Spearmint (*Mentha spicata* L), belongs to the family Lamiaceae. The plants of this family are a rich source of polyphenols and thus possessing strong antioxidant properties. Spearmint is native to northern England and is essentially grown in areas with climate ranging from tropical to temperate, such as America, Europe, China, South Africa and Brazil. Nowadays, spearmint is widely grown throughout all regions in the world.

HABITAT

Mentha spicata grows in moist places such as along shorelines and stream banks. This herb can grow in low elevations upward into the higher elevations in the mountains.

Medicinal Properties

Spearmint is a normally used as a domestic herbal remedy. A tea made from the leaves has conventionally been used in the treatment of fevers, headaches, digestive disorders and various minor ailments. The herb is antiemetic, antispasmodic, carminative, diuretic, restorative, stimulant and stomachic. The leaves should be harvested when the plant is just coming into flower, and can be dried for later use. The stems are macerated and used as a poultice on bruises. The essential oil in the leaves is antiseptic, though it is toxic in large doses. Both the essential oil and the stems are used in folk remedies for cancer. A poultice prepared from the leaves is said to remedy tumours.

Other Uses

An essential oil is obtained from the whole plant, the yield being about 4 Kg of oil from 1 tonne of leaves. The oil is used commercially as a food flavouring and oral hygiene preparation. Rats, mice and a few insects intensely dislike the smell of mint. The plant was therefore used in homes as a strewing herb and has also been spread in granaries to keep the rodents off the grain.

Botany

Spearmint is a rhizomatous, upright perennial most commonly grown as a culinary herb. The trademark of mint family is the stem which is in square-shaped. The leaves are broad and sharply serrate. It typically grows to 0.6-1 m and spreads by rhizomes to form an attractive ground cover. Spearmint produces rightly and long pink or white flowers in slender spikes. Spearmint leaves possess a characteristic aromatic odour and pungent taste. However, in contrast to peppermint and Japanese mint, the cooling after-effect is absent in spearmint. It flowers from August to September, and the seeds ripen from September to October. The species is hermaphrodite (has both male and female organs) and is pollinated by bees.

Chemical Constituents

Carvone, a phenolic compound, is the main constituent found in spearmint oil, followed by limonene. It inhibits bacterial growth, act as fungicide and as insect-repellent as well. Carvone also reversibly suppresses the sprouting in stored potatoes or flower bulbs. Carvone renders the characteristic smell of spearmint. The smell is smooth and near analogous to fennel oil. However, (S)-carvone and (R)- carvone contribute to the odor of spearmint and caraway seeds respectively. Both are enantiomers of a biological material but elicit different response. In addition, S-carvone possesses high antioxidant activity.

Agrotechniques

Loam and sandy loam soil having a pH between 6.0-7.5 with good irrigation facilities are ideal for its cultivation. It can grow in semi-shade or no shade. Cuttings are usually planted at a spacing of 45 x 20 cm. At least 4-5 times ploughing is required in the land where FYM or compost fertilizers are added with the soil before ploughing. Nitrogen, phosphorous and potash can be applied as and when required. After application of fertilizer, irrigation and weeding is needed. Spearmint is in bloom from August to September. The plant needs frequent watering so that the soil is always moist. It can be a good companion plant to cabbages and tomatoes because it wards of pests and insects. This helps lower the need for pesticides.

24. Common name: Bon peyaj, Welsh onion, Bunching onion, Spring onion

Scientific name: *Allium fistulosum/ Allium cepa* L.

Family: Amaryllidaceae

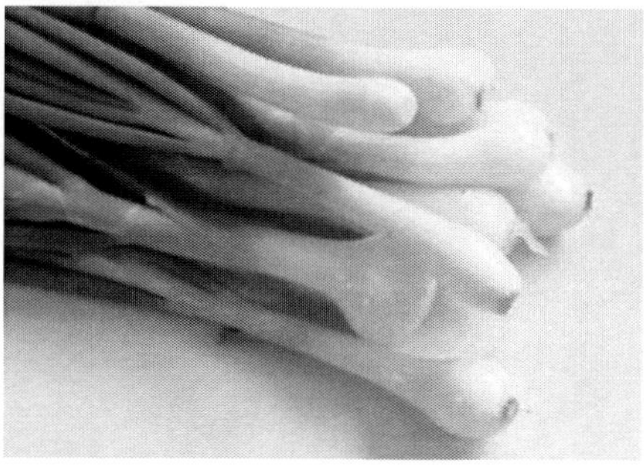

INTRODUCTION

The Japanese bunching onion, often called the welsh onion in Britain, is widely cultivated in eastern Asia from the cold temperate regions of Siberia to moderate elevations in the tropics of southeast Asia and commonly sold in shops and markets. A very common vegetable in Asia, where there are many different cultivars, it is usually grown more on a garden scale in Europe and America, mainly for its edible leaves which can be produced throughout the winter if the weather is not too severe.

HABITAT

Anthropogenic (man-made or disturbed habitats), meadows and fields.

MEDICINAL PROPERTIES

It is used to improve the functioning of internal organs and the metabolism, and to prolong life. It improves eyesight, aids in digestion and perspiration, and enhances recovery from common colds, headaches, wounds and festering sores. The bulb contains an essential oil that is rich in sulphur compounds. It is antibacterial, antiseptic, diaphoretic, diuretic, galactagogue, stomachic and vermifuge. A tea made from the roots is a children's sedative. Use of the bulb in the diet impedes internal parasites. Externally, the bulb can be made into a poultice to drain pus from sores, boils and abscesses.

OTHER USES

Leaf is used as a leafy vegetable. The juice of the plant is used as a moth repellent. The whole plant is said to repel insects and moles. The plants are said to reduce or prevent white ant infestation in gardens, and in China the diluted juice is used as a treatment against aphids. The welsh onion is related to the cultivated onion (*Allium cepa*) and could be of value in breeding programmes.

BOTANY

Pungency and smell of bulb is as like as onion. *Allium fistulosum* is a usually evergreen, herbaceous, perennial plant producing 2 - 6 hollow, cylindrical leaves 25 – 40 cm long and a flowering scape 30 – 50 cm tall from an underground bulb. The plant divides quite freely, forming a cluster of growth.

CHEMICAL CONSTITUENTS

A.cepa is known to contain a variety of sulphur-containing compound, also responsible for the characteristic pungent taste and lachrymatory factor. The inhibitory effects of sulphur-containing compounds found in fresh and/or cooked onion on the differentiation of white adipose cells has been recently reported.

A nutritional analysis revealed that they contain about 1.4% protein, 0.3% fat, 4.6% carbohydrate, 0.8% ash, some vitamin B1 and moderate levels of vitamin C. The bulb contains an essential oil that is rich in sulphur compounds.

AGROTECHNIQUES

Allium fistulosum is a very cold-tolerant plant that originated in the temperate zone and is widely grown there. Plants will often retain their leaves even when covered in snow. In the tropics, it can be grown successfully at elevations between 1,000 - 2000 metres. High temperatures support bulb formation, but flower formation and seed production are only possible where the bulbs are subjected to low temperatures. In the tropics, flower and seed formation will, therefore, only occur at higher elevations. The plant grows best in areas where annual daytime temperatures are within the range 12 - 25°C. It prefers a mean annual rainfall in the range 850 - 1,600mm. It prefers a sunny position in a light well-drained moist soil, but tolerates most soils, including those that are damp and acidic. It essentially prefers a pH in the range 6.5 to 7.5. Average yields of leaves may be 34-45 t/ha. The plants are often eaten by slugs.

25. Common name: Brahmi, Water hyssop, Herb of grace
Scientific name: *Bacopa monnieri*
Family: Scrophulariaceae

INTRODUCTION

It is mainly found in warmer and humid conditions. It is a creeping herb with average height of 2-3 feet and branched roots at the nodes. Flowers are of whitish or pale blue in color with small and oval shaped fruit. Seeds are 0.2-0.3 mm in size with dark brown in color. North and South America, Europe, Australia, southern India, Asia and Africa are major countries growing Brahmi.

HABITAT

This creeping herb is found to grow along the canals and water bodies and in marshy areas. Bramhi can also be grown in normal places like near water pump, pond and swampy places.

MEDICINAL PROPERTIES

Whole herb i.e., seeds, roots, leaves, rhizomes etc. are used for preparing various drugs. Drugs prepared from Brahmi is used as anticancer and used to cure anemia, asthma, diuretic, tumours and epilepsy. It is also used as an antidote for snake bite. The whole plant is used in indigenous system of medicine as a nerve tonic and for epilepsy and insanity. It is digestive and can be used by extracting juice or frying It is also being used as diuretic and for treating rheumatisim, asthma and hoarseness. Besides, brahmi also has good potency in controlling cough, fever and diabetes. Because of its inherent potential of enhancing memory and strength, this wonder plant is getting hold of attention for its commercial cultivation worldwide.

The plant is considered among one of the 'celestial drugs' (*Divya ausadhi*), when consumed with milk for six months. In Siddha system of medicine, the plant is useful against painful joints, swelling in joints, peripheral neuritis, constipation and burning urination. It is also used in convulsion, mental retardation, chest congestion and laryngitis.

Brahmi is a reputed intellect promoter, hypotensive, and neuropathic sedative drug. It is a potent tonic used to improve mental alertness, learning performance, and memory as well as in the treatment of insomnia. It protects from arthritic problem. It is the principal ingredient of classical Ayurvedic preparations like brahmighritam and brahmirasayanam.

Other Uses

Bacopa monnieri has been used by Ayurvedic medical practitioners for centuries for a variety of purposes, including improving memory, reducing anxiety, and treating epilepsy. In fact, research shows that it may boost brain function and alleviate anxiety and stress, among other benefits.

Botany

Bacopa is a succulent, glabrous, creeping herb, with rooting at nodes. The plant is easily recognized by its spreading habit, sessile and fleshy leaves, and light bluish, purple or white flowers. Leaves are ovate and opposite with dotted lower surface. Leaf sessile and fleshy leaves, and light bluish, purple or white flowers. Leaves are ovate and opposite with dotted lower surface. Rooting occurrs in nodes of stem.

CHEMICAL CONSTITUENTS

The main nootropic constituents of *Bacopa* are believed to be dammarane types of triterpenoid saponins known as bacosides, with jujubogenin or pseudo-jujubogenin moieties as aglycone units. Bacosides comprise a family of 12 known analogs. Novel saponins are called bacopasides I–XII that have been identified more recently. The alkaloids brahmine, nicotine, and herpestine have been catalogued, along with D-mannitol, apigenin, hersaponin, monnierasides I–III, cucurbitacins and plantainoside B.One of the most studied constituents has been bacoside A, which is found to be a blend of bacoside A_3, bacopacide II, bacopasaponin C, and a jujubogenin isomer of bacosaponin C.These assays have been conducted using whole plant extract, and bacoside concentrations may vary depending upon the part from which they are extracted.

AGROTECHNIQUES

The plant occurs in a variety of soil types if the habitat provides wet and semi-shade conditions. It is found at altitudes up to 1300 m. In North India, it can grow in a wide range of temperatures 15–40 °C and soil pH ranging 5–7.5. It is a plant that grows under sub-tropical and tropical climate. A temperature range of $33^0 – 40^0$C with relative humidity of 60-65 percent has been found for its optimum vegetative growth. Near-neutral, clayey loam to clayey soils are best suited for the growth of *Bacopa monnieri*. However, it can even grow well in soils with pH of 7.5 or even more. It becomes dormant during the winter months except when grown near running water. The plant grows well in poorly drained soils.

Cuttings are used as a planting material. For brahmi plantation, ideal time of planting is April and May. It required well pulverize and levelled soil. To bring soil to fine tilth, ploughing and harrowing is to be done. Irrigation is done immediately when land is converted to plots. Application of FYM @20q/acre is mixed with soil at the time of ploughing. The planting should be done in the month of mid-June or early July. Regarding the inorganic fertilizers application of 100:60:60kg/ha of NPK is recommended. Seedlings are transplanted at the distance of 20 x 20cm

It is essential to water the field after transplanting for the survival and establishment of the plants. Subsequently, the field should be irrigated at 7-8 days interval. Irrigation should be avoided during rainy season. Primarily, hand weeding is needed at every 15-20 days interval but later on as plants proliferate and form a dense mat of vegetation, weeding may be done occasionally.

26. Common name: Bathua, Lamb's quarters, Goosefoot, Wild spinach

Scientific name: *Chenopodium album*

Family: Amaranthaceae

INTRODUCTION

Chenopodium album is an erect, annual plant usually growing up to 150 cm tall, though when growing in rich ground and in the longer day lengths in the temperate zone, it can sometimes reach up to 400 cm. The plant is harvested from the wild for local use as a food and medicine and, in some areas, is quite popular. A common weed of cultivated ground, it is also occasionally cultivated as a food crop.

HABITAT

Chenopodium album is one of the most widely distributed species of weeds in the world and is one of the most successful colonizers as it moves into new areas. Being a colonizing species it occurs on habitats that have been opened up by disturbances. Better adapted to temperate zones, *Chenopodium album* is a secondary weed in the tropics.

The *Chenopodium album* grows among the crops that have been dragged, spring grains, among the ruins, the rivers, the soils rich in nitrogen more or less saturated. It grows between 0 and 1500 m above mean sea level.

MEDICINAL PROPERTIES

The leaves are anthelmintic, antiphlogistic, antirheumatic, mildly laxative, odontalgic. An infusion is taken in the treatment of rheumatism. The leaves are applied as a wash or poultice to bug bites, sunstroke, rheumatic joints and swollen feet, whilst a decoction is used for carious teeth. The seeds are chewed in the treatment of urinary problems and are considered useful for relieving the discharge of semen through the urine. The stem juice is applied to freckles and sunburn. The root juice is used in the treatment of bloody dysentery. Food that comprises 25.5% of the powdered herb may suppress the oestrus cycle.

OTHER USES

Leaves are consumed raw or cooked. Edible seeds are dried and ground into a meal and eaten raw or baked into bread. The seed can also be sprouted and added to salads. The seed is very difficult to harvest and to use due to its small size. The seed should be soaked in water overnight and thoroughly rinsed before being used in order to remove any saponins. Young inflorescences are also cooked as a tasty broccoli substitute.

BOTANY

A lambs quarter is an annual species, always propagated from seeds. Each plant produces about 30 to 40000 seeds, but large plants may produce 500000 seeds. The seeds may be dispersed by birds, or transported by the water. Seeds have been known to survive from 30 to 40 years in soil. Only the seeds in superficial layers of the soil are able to germinate. There are two types of seeds, some with hard envelope and dormant, and others with a supple envelope that can germinate immediately.

CHEMICAL CONSTITUENTS

The widespread uses of *Chenopodium* genus in traditional medicine have resulted in considerable chemical analysis of the plants and their active principles. The phytochemical investigations of genus *Chenopodium* have afforded compounds with vast variety of structural patterns. From the phytochemical point of view, the chenopods were reported to contain: minerals, primary metabolites viz., carbohydrates, amino acids, nonpolar constituents, proteins, aromatic cytokinins, hormones and secondary metabolites like flavonoids, saponins,terpenes,sterols,alkaloids and vitamins. The leaves contain about 3.9% protein, 0.76% fat, 8.93% carbohydrate, 3% ash. The seed contains about 49% carbohydrate, 16% protein, 7% ash, 5.88% ash.

AGROTECHNIQUES

The plant can be found in most regions of the world, from cold temperate to the tropics especially at higher elevations. It has a broad tolerance of climates, succeeding in areas with average temperatures ranging from 5 - 30°C, and is tolerant of night frosts. It is an easily grown plant, succeeding in most soils but disliking shade. It prefers a moderately fertile soil. It can tolerate a wide range of pH between 4.5 to 8.3.The plant is a common weed of disturbed ground through much of the world. In moderate amounts this plant is a good companion for potatoes, corn and cucurbits. The plant responds directly to the magnesium content of the soil so it can be used to indicate the presence of that element.

Each chenopod plant can produce tens of thousands of seeds. The chenopod is very prolific in the production of seeds. From each plant, about 60,000 seeds can be obtained. In addition, these seeds last very long, up to 50 years. The drying of seed is completed in a sheltered place, on dry paper or paperboard. Then the seeds are separated from the rest of the plant, even though it is not necessary to eliminate all the glomeruli because they are also edible.

27. Common name: Biliti dhonepata, Mexican/Long/Sawtooth coriander

Scientific name: *Eringium foetidum*

Family: Apiaceae

Introduction

It is sometimes cultivated as a coriander-like flavouring in the tropics. The plant is harvested from the wild for local use as a food and medicine.

Habitat

Open rocky places at elevations of 700 - 1,200 metres where it is naturalised in Nepal. Meadows, plantations, waste places, along roadsides and forest edges, from lowland areas up to 1,700 metres. Also found in swampy area with less care and management.

Medicinal Properties

A decoction of the whole plant is used as an anti-malarial and for the treatment of haemorrhages. A decoction of the crushed leaves is used as a treatment for children's leprosy and children's convulsions. The plant is boiled and the water used for a herbal bath or as a medication for chicken pox and measles. The leaves and roots are boiled and the water drunk for pneumonia, flu, diabetes, constipation, and malaria fever. The leaves are febrifuge, laxative. The root is abortifacient, febrifuge, stomachic, sudorific. An infusion is used to treat hydropsy and stomach pains. The root can be eaten raw for scorpion stings. In India, the root is reportedly used to alleviate stomach pains. The leaves as such can be consumed in the form of chutney as an appetite stimulant.

Other Uses

The fresh leaves are used as a flavouring in food, e.g., in soups, curries, stews, rice and fish dishes. Tender young leaves are eaten raw or cooked, as a vegetable. They can be steamed and served with rice. The root is used as flavouring in soups. Although the raw root has an offensive odour, it imparts a very pleasant flavour to the food.

Botany

It is an erect, perennial spiny herb growing to a height of 20-80 centimeters. Stems are much-branched from a stout taproot. Leaves are in basal rosette. Leaf blade is10-16 centimeters long, 3-4 centimeters wide, narrow, with rounded tips, tapered base, and toothed margins. Flowers are pale green, minute, in dense conical heads, 1 centimeter long, above a ring of spiny bracts. Fruit is 0.2 centimeter, spiny, and 2-seeded.

Chemical Constituents

Study of various parts of *Eringium foetidum* for essential oil by GC-MS analysis yielded dominant constituents viz.,(E)-2-Dodecenal (28.43%), 13-tetradecenal (27.45%), dodecanal (14.59%) and 2,4,5-trimethyl-benzaldehyde (10.77%) in leaf;dodecanal (20.21%), 2,4,5-

trimethylbenzaldehyde (18.43%) in stem;and2,4,5-trimethylbenzaldehyde (56.08%), 13-tetradecenal (9.26%) and (E)-2-dodecenal (7.65% in root. GC-MS analysis of seed of *E. foetidum* of Cuban origin yielded 37 compounds. Major constituents were carotol (19–31%), (E)-β-farnesene (9.98%), (E)-anethole (7.43%) and α-pinene (7.69%).

AGROTECHNIQUES

It prefers a well-drained soil with abundant sunlight. Sandy loam is better for its cultivation. A partly shaded site and a fertile not too- dry soil is desirable. It can be cultivated in pot also. Planting material is seed. Plants do not usually grow close together but emerge as individual weeds. May and June months are ideal for its cultivation. It can be cultivated in October month also. However, in some parts of the world, cilantro can be cultivated throughout the year. Plants can flower and produce seed all year round. Application of fertilizer and irrigation in proper time can increase its production. Individual tender leaves are harvested, leaving the topmost 3 leaves intact. On an average, 10 - 15 leaves can be harvested per plant in 5 - 10 harvests at intervals of 1 - 2 weeks before flowering intervenes. Leaves wilt easily and so are best harvested just before use. After harvesting leaf can be use for seven days by applying water to the leaf.

28. Common name: Bokful, sesbania, Vegetable Hummingbird
Scientific name: *Sesbania grandiflora*
Family: Fabaceae

Introduction

Sesbania grandiflora is a short-lived, soft-wooded, loosely-branching tree with a rather open crown. It is a true multi-purpose tree, providing a range of foods, medicines, timber, gum and tannins, mainly for local use. The plant is cultivated in many tropical areas as an ornamental, a green manure crop, in soil reclamation schemes and for its many useful properties. It is sold as a vegetable in local markets.

Habitat

Dry wasteland and roadsides

Medicinal Properties

The leaves are aperient and diuretic. Crushed leaves are applied as a poultice to sprains and bruises of all kinds, swellings, rheumatism, itching etc. A tea made from the leaves is antibiotic, anthelmintic, antitumor and contraceptive. The bitter bark is astringent, febrifuge, tonic and antipyretic. It has a remedy for gastric troubles, colic with diarrhoea and dysentery. A bark decoction is taken orally to treat fever, diarrhoea, dysentery and diabetes. The flowers are emollient and laxative. Juice of the flowers, put in the eyes, is said to relieve dimness of vision. Sinus congestion is reduced by taking a decoction of the flowers. The root is a well-known medicine for malaria. Root juices are used for poultices. A paste of the root is applied externally in the treatment of rheumatism. Juice of the root, mixed with honey is used as an expectorant. Decoction of the bark is used for hemoptysis. Infusion of the bark is given for smallpox and other eruptive fevers. Flower extracts are used to treat wrinkles as well as leaves as poultice for bruises. Decoction of bark is used as vomitive. In the Antiles, bitter bark is tonic and febrifuge. In Ayurveda, fruits are used for anemia, bronchitis, fever, tumors; flowers for gout, bronchitis, nyctalopia. Juice of leaves considered anthelmintic and tonic, and used for biliousness, itching, and night blindness.

OTHER USES

Flowers are often consumed as raw or cooked. Added to salads, they are boiled as a potherb, fried or used in curries. It is rich in sugar and iron, with a flavour like mushrooms. The centre part of the flower is usually removed because it is very bitter. White flowers are generally preferred to red ones. The long, narrow pods are boiled and eaten like string beans. The very young seedpods can be added to salads. The pods can be up to 50cm long and 8mm wide. The fruits, falling leaflets and flowers make excellent green manure or mulch and improve soil fertility.

BOTANY

Sesbania grandiflora is a fast-growing perennial, deciduous or evergreen legume tree, up to 10-15 m high. Its lifespan is about 20 years. Its roots are heavily nodulated and some floating roots may develop in waterlogged conditions. The trunk is straight with few branches. The leaves, up to 30 cm long, are pinnately compound with 20-50 oblong leaflets, 1-4 cm long and 0.5-1.5 cm broad. The flowers are white, yellowish, pink or red and borne in axillary racemes. The pods are 50-60 cm long, glabrous and indehiscent, and hang vertically. They contain 15 to 50 dark brown seeds, 5 mm long and 2.5-3 mm broad.

Chemical Constituents

Bark contains tannin and gum. Saponin is isolated from the seeds. Sesbanimide has been isolated from seed is considered a cancer inhibitor. Flower yields proteins, tannins, oleanolic acid, kaempferol, cystine, isoleucine, aspargine, phenylalanine, valine, nicotinic acid, vitamin C. Phytochemical screening of aqueous extracts yielded carbohydrates, proteins, flavonoids, alkaloids, tannins, and glycosides. Phytochemical screening of flowers yielded carbohydrates, proteins, amino acids, saponins, flavonoids, alkaloids, tannins, and glycosides. Study on roots isolated three isoflavanoids, isovestitol, medicarpinand sativan, along with a known compound, betulinic acid. Study of stem bark of *S. grandiflora* isolated two new 2-arylbenzofurans, sesbagrandiflorain A and B. Nutrient analysis of raw Sesbania flower per 100 g is:

- Proximates- water 91.58 g, energy 27 kcal, total fat/lipid 0.04 g, carbohydrate by difference 0.73 g;
- Minerals- calcium 19 mg, iron 0.83 mg, magnesium 12 mg, phosphorus 30 mg, potassium 184 mg, sodium 15 mg;
- Vitamins- vitamin C 73.0 mg, thiamin 0.083 mg, riboflavin 0.081 mg, niacin 0.430 mg, folate 102 μg;
- Lipids- trans fatty acid, cholesterol.

AGROTECHNIQUES

Sesbania grandiflora is well adapted to hot, humid environments in the lowland tropics. It succeeds at elevations below 1,000 metres with a mean annual temperature in the range 22 - 30°c, and a mean annual rainfall of 2,000 - 4,000 mm, exceptionally tolerating rainfall as low as 800mm per year. It lacks tolerance for cool temperatures below about 10°c. It seems to prefer a bimodal rainfall distribution, growing rapidly during the wet season, but is capable of withstanding prolonged dry seasons of up to 9 months.

It requires a sunny position for best growth. It prefers a fertile, moist but well-drained moderately light soil, though it succeeds in light sandy, medium, heavy clayey and low fertility soils. Prefers a pH in the range 5.5 - 8.5, but can tolerate acid conditions down to 4.5. Requires a position sheltered from strong winds. The plant has an outstanding ability to tolerate water logging and is ideally suited to seasonally flooded environments. When flooded, the plant initiates floating, adventitious roots, and protects its stems.

Widely cultivated in the tropics as an ornamental, it is valued especially for its long, flowering display. The plant is a prolific producer of seeds, leading to a lot of litter under the tree and a tendency to produce seedlings in unwanted places. It has escaped from cultivation in some areas and become naturalized. It is considered to be invasive in some Pacific islands.

The species is very fast growing, but does not live long - it has a life span of about 20 years. Plants have reached a height of 3.2 metres in just 9 months when raised in loamy soils, but only 1.8 metres when growing in sandy soils. The plant is shallow-rooted and tends to grow too quickly, resulting in brittle and weak limbs that have a tendency to snap and a plant that is easily uprooted in strong winds.

At a very short rotation of 3 - 4 years, the tree is capable of producing much higher cellulose raw material per unit area than most other pulp woods. Even trees of 3 - 4 years old can be pulped without debarking and are suitable for chemical pulping for use as cheap printing, writing, magazine and newsprint paper, etc. On a 3-year rotation, about 41 tonnes of pulp per hectare per year can be harvested.

29. Common name: Misti alu, Sweet potato, Spanish potato
Scientific name: *Ipomea batatas*
Family: Convolvulaceae

INTRODUCTION

Sweet potato is a vigorous scrambling or climbing perennial plant producing stems 4 metres or more long from underground tubers. Leaf, steam of sweet potato are used as a vegetable and also used as a fodder crop by boiling it. The plant was developed in cultivation, probably from the Central American species *Ipomoea trifida*. It is one of the most widely grown root crops in the tropics, where it is a staple food. It is also one of only seven world food crops with an annual production of more than 100 million metric tons. The root is also used as a source of an industrial starch.

Habitat

It can adapt in a wide range of different environments. This plant can grow from cool temperate grassland to warm wet tropical climates.

Medicinal Properties

The leaf decoction is used in folk remedies for tumors of the mouth and throat. It is reported to be alterative, aphrodisiac, astringent, bactericide, demulcent, fungicide, laxative, and tonic, sweetpotato is a folk remedy for asthma, bugbites, burns, catarrh, ciguatera, convalescence, diarrhea, dyslactea, fever, nausea, renosis, splenosis, stomach distress, tumors, and whitlows. Polyphenols have physiologic functions, radical scavenging activity, antimutagenic, anticancer, antidiabetes and antibacterial activity in vitro and vivo.

Other Uses

Sweet and fleshy roots are cooked as a delicious staple food and are also very nutritious providing a rich source of vitamins and minerals. The crop is cultivated mainly for the tuber, used as vegetable, either boiled, baked, fried, or dried and ground into flour to make biscuits, bread, and other pastries. Tubers are also dehydrated in chips, canned, cooked and frozen, creamed and used as pie fillings. Leafy tops are eaten as vegetable and sold in markets in Malaysia. It is greatly admired as feed for farm animals. The haulms or straw after threshing are used as green manure.

BOTANY

It is a spreading, prostrate, herbaceous, smooth or somewhat hairy vine. Stems or runners are several meters long, take root when in contact with soil. Leaves are ovate to oblong-ovate, 6 to 14 centimeters long, somewhat entire, with alternate heart-shaped or palmately lobed leaves, pointed at the tip, and heart-shaped at the base. Flowers are sympetalous, 5 centimeters long, funnel-shaped, purple, self sterile, and rarely producing seeds. Stamens are 5, of different lengths, epipetalous, with hairy filaments. Sepals are about 1 centimeter long and greenish with pointed lobes. Corolla is funnel shaped, 4 to 5 centimeters long, pink, purple, and whitish. The edible and sweet-tasting tuberous root is long and tapered with a smooth skin of varying colors from red, purple, brown and white. The flesh varies from white, orange, purple and yellow.

CHEMICAL CONSTITUENTS

Leaves have a high content of polyphenolics - anthocyanins and phenolic acids, with at least 15 biologically active anthocyanins. Study isolated four compounds: citrusin C, caffeic acid, 3,4-di-O-caffeoylquinic acid and 1,2,3,4-tetrahydro-beta-carboline-3-carboylic acid.Nutrient and anti-nutrient study of leaves yielded: crude protein, 24.85%; crude fat, 4.90%; crude fibre, 7.20%; ash, 11.10%; carbohydrate, 51.95%; moisture contents, 82.21%; and calorific values, 351.30. Vitamin composition showed: vitamin A, 0.672 mg/100g; vitamin C, 15.20 mg/100g. Elemental analysis showed appreciable levels of zinc (0.08), potassium (4.05), sodium (4.23), manganese (4.64), calcium (28.44), magnesium (340.00) and iron (16.00). It showed low levels of toxicants. (See studies below).Nutrient value of sweet potato per 100 g yields energy 360 kJ (86kcal), starch 12.7 g, sugars 4.2 g, dietary fiber 3.0 g, fat 01 g, protein 1.6 g, vitamin A equivalent 709 μg, thiamine (B1) 0.1 mg, riboflavin (B2) 0.1 mg, niacin (B3) 0.61 mg, pantpthenic acid (B4) 0.8 mg, pyridoxine

(B60 0.2 mg, folate (B9) 11μg, vitamin C 2.4 mg, vit E 0.26 mg, calcium 30.0 mg, iron 0.6 mg, magnesium 25.0 mg, phosphorus 47.0 mg, potassium 337 mg, sodium 55 mg, zinc 0.3 mg.

AGROTECHNIQUES

Seeds are rarely used for propagation purposes, all propagation is by vegetative means.Sweetpotatoes should be transplanted only after the soil has warmed up well, in late spring. It is undesirable to grow sweet potatoes on soils too rich in manures. Early transplanting is best. Soil fertility and freedom from diseases must be maintained. Best cuttings consist of a length of 20 cm. Plants are spaced about 30 cm apart in rows about 1.1 m apart (31,250 plants/ha), or in rows 1.3 m apart (26,350 plants/ha), or in rows 1.6 m apart (21,250 plants/ha). High yields of high quality sweet potatoes can be obtained only on soils of high fertility, good drainage, and suitable texture. Plants respond well to good soil management and fertilizer practices. The most profitable yields have been generally obtained with 1.5–2.25 MT of fertilizer mixtures containing 2–4% nitrogen, 8% phosphoric acid, and 8–10% potash. On fields where good amounts of green manures or animal manures have been turned under, often no more than 2% of nitrogen is needed. On lighter sandier soils, 10% of potash is recommended. Fertilizer generally should be applied in two stages, one just before the ridges are prepared ca 2 weeks before planting, when 0.75 tonnes/ha is thoroughly mixed in the soil where the ridge is to be, with the ridge thrown up so that most of tile fertilizer lies deep in the ridge, and the remainder applied as top-dressing to the sides of the ridge 2 or 3 weeks after transplanting. In general, on furrow-irrigated lands, six to eight irrigations are applied between transplanting and harvesting. In all cultivations, weed control must be pratically perfect before the vines become long enough to interfere with cultivation. Sweet potatoes are poor competitors with weeds. Weeds reduce yields and interfere with harvesting. In order for the roots to store through the winter, they need to be cured in the sunshine at temperatures around 25°C for about a week before being stored at around 14°C.

30. Common name: Methi, Greek-clover, Greek hay
Scientific name: *Trigonella foenum-graecum*
Family: Fabaceae

INTRODUCTION

Fenugreek (*Trigonella foenum-graecum* L.) is an annual herbaceous legume suitable for dry land areas where moisture is not sufficient. Fenugreek plant and seeds have a characteristic strong odour. The seeds are used as condiments. Fenugreek is sometimes used as a short-rotation catch crop after sugarcane or cotton. Native of the waste grounds of the Mediterranean basin, North-eastern Africa and Western Asia, it has been cultivated since ancient times and is now days widely cultivated in India, France and Turkey.

HABITAT

Field verges, uncultivated ground, dry grasslands and hillsides.

Medicinal Properties

Fenugreek is an acclaimed herbal medicine, especially in North Africa, the Middle East and India. It has an extensive assortment of medicinal applications. The seeds are very much beneficial and are prearranged to convalescents and to persuade weight gain, particularly in anorexia nervosa. Seeds can inhibit cancer of the liver, lower blood cholesterol levels and also have an antidiabetic effect. The seed and leaves are anticholesterolemic, anti-inflammatory, antitumor, carminative, demulcent, deobstruent, emollient, expectorant, febrifuge, galactogogue, hypoglycaemic, laxative, parasiticide, restorative and uterine tonic. The seed is very nourishing and is one of the most efficacious tonics in cases of physical debility caused by anaemia or by infectious diseases. It is also used in the treatment of late-onset diabetes, poor digestion, insufficient lactation, painful menstruation, labour pains etc. The seeds clean up bad breath and reinstate a blunted sense of taste. Externally, the seeds can be ground into a powder and used as a poultice for abscesses, boils, ulcers, burns etc. Compounds extracted from the plant have shown cardiotonic, hypoglycaemic, diuretic, antiphlogistic and hypotensive activity.

Other Uses

Fenugreek is used as an herb (dried or fresh leaves), spice (seeds), and vegetable (fresh leaves, sprouts, and microgreens). Sotolon is the chemical responsible for the distinctive maple syrup smell of fenugreek. Cuboid-shaped, yellow- to amber-coloured fenugreek seeds are commonly seen in the cuisines of the Indian subcontinent, used both whole and powdered in the preparation of pickles, vegetable dishes, dal, and spice mixes such as panch phoron and sambar powder. They are often roasted to reduce inherent bitterness and to enhance flavour.

BOTANY

Trigonella foenum-graecum is an aromatic annual growing up to 60 cm. Fenugreek is an erect, smooth, herbaceous plant that can grow up to a height of 40-80 cm. It is taprooted. Its stems are erect, up to 50 cm high, sometimes branched. The leaves are alternate, compound, obovate, trifoliolate, 7-12 cm long, light green in colour. The flowers are pale yellow in leaf axil. The pods are curved and sickle-shaped, and there are 10 to 20 angular seeds per pod which are aromatic, yellow-brown, rhomboidal and divided into two lobes by a furrow.

CHEMICAL CONSTITUENTS

The major components are N-methylhomopiperazine, 8-nonynoic acid, 1-octanol, 2,7-dimethyl-,2,4-Pentanedione, 3-(2-propenyl), Undecanal, 2-methyl, 1-Iodo-2- methylundecane, didodecyl phthalate, vitamin E, 5à-Androstan-16-one, cyclic ethylene mercaptole, ccholestan-3-ol, 2-methylene-, (3á,5à), 2H-pyran, 2-(7- heptadecynyloxy) tetrahydro-, and cholest-5-ene, 3-bromo-, (3á) compounds. These phytochemicals are responsible for various pharmacological actions like antimicrobial, antioxidant, and anti-inflammation, anti cancer activities.

AGROTECHNIQUES

Fenugreek grows best in areas where annual daytime temperatures are within the range 13 - 27°C. Plants are hardy to about -15°c. It prefers a mean annual rainfall in the range 600 - 800mm. It succeeds in ordinary garden soil, preferring a well-drained loamy soil in full sun with a pH in the range 7 - 7.5. Pre-soaking the seed for 12 hours in warm water is beneficial before sowing in situ. Plants take about 16 weeks to mature their seed in warmer climates, probably about 4 weeks longer in temperate areas. The seed is ripened intermittently over a period of some weeks making harvesting more complicated. The seeds are harvested when fully ripe and dried for later use. The expected yield of seed is 0.5 - 3.8 t/ha. In India, yields of green forage are about 9 - 10 t/hatches species has a symbiotic relationship with certain soil bacteria. These bacteria form nodules on the roots and fix atmospheric nitrogen. Some of this nitrogen is utilized by the growing plant but some can also be used by other plants growing nearby. When removing plant remains at the end of the growing season, it is best to only remove the aerial parts of the plant, leaving the roots in the ground to decay and release their nitrogen.

31. Common name: Mulo, daikon
Scientific name: *Raphanus sativus L*
Family: Brassicaceae

INTRODUCTION

Radish is an erect, annual plant, generally in the region of 20 – 100 cm when flowering. It grows from a globose to cylindrical root. The plant is extensively cultivated throughout both tropical and temperate regions for its edible roots. While this plant is often grown in vegetable gardens, it is uncommon to find the non-native Garden Radish in the wild.

HABITAT

Habitats include dumps, edges of gardens, areas along roadsides and railroads, and waste areas. The preference is disturbed habitats, where this plant usually doesn't persist.

MEDICINAL PROPERTIES

Radish roots stimulate the appetite and digestion, having a tonic and laxative effect upon the intestines and indirectly stimulating the flow of bile. The plant is used in the treatment of intestinal parasites, though the part of the plant used is not specified. The leaves, seeds and old roots are used in the treatment of asthma and other chest complaints. The juice of the fresh leaves is diuretic and laxative. The seed is carminative, diuretic, expectorant, laxative and stomachic. It is taken internally in the treatment of indigestion, abdominal bloating, wind, acid regurgitation, diarrhoea and bronchitis. The root is antiscorbutic, antispasmodic, astringent, cholagogue, digestive and diuretic. It is crushed and used as a poultice for burns, bruises and smelly feet. Its use is not recommended if the stomach or intestines are inflamed. The plant contains raphanin, which is antibacterial and antifungal. It inhibits the growth of *Staphylococcus aureus*, *E. coli*, streptococci, pneumococci etc. The plant also shows anti-tumour activity.

OTHER USES

The growing plant repels beetles from tomatoes and cucumbers. It is also useful for repelling various other insect pests such as carrot root fly. Radishes are a good companion plant for lettuces, nasturtiums, peas and chervil, tomatoes and cucumbers. They are said to repel cucumber beetles if planted near cucumber plants and they also repel the vine borers which attack squashes, marrows and courgettes. They grow badly with hyssop and with grape vines.

There is a fodder variety that grows more vigorously and is used as a green manure. A non-drying oil is obtained from the seeds. It can be used in soap making.

BOTANY

This annual or biennial plant consists of a rosette of leaves; afterwards it bolts and produces flowering stems up to 80 cm tall. The basal leaves are up to 15-20 cm long and 6-7 cm across; they are oblanceolate, coarsely crenate, and pinnately lobed. These lobes may be shallow or deep; the terminal lobe is always the largest. The basal leaf surface is usually rough and with stiff hairs. The central stem is often reddish at the base, but light green elsewhere; it is either glabrous or covered with scattered stiff hairs. The upper side stems are very similar, except that there is often a red ring where they branch from the central stem. The alternate leaves on the stems are similar in appearance to the basal leaves, except that they are smaller, less likely to be deeply lobed, and narrowly ovate in shape.

CHEMICAL CONSTITUENTS

Alkaloid and nitrogen compounds present in the roots are pyrrolidine, phenethylamine, N methylphenethylamine, 1, 2´-pyrrolidin-tion-3-il-3-acid-carboxilic-1, 2, 3, 4-tetrahydro- β- carboline, and sinapine. Cytokinin (6-benzylamino-9-glucosylpurine) is a major metabolite of 6-benzylaminopurine (6-BAP) in the root radish. A minor metabolite of 6-BAP from radish has been identified as 6-benzylamino-3-β-D-glucopyranosylpurine. Total amino acids are 0.5% of dry wt; with proline (0.5%) as the major constituent, methionine and cystine are present in traces (0.02%). Diamines as diaminotoluene (2,4-D), 4,4´-

methylenedianiline (4,4-D), and 1,6-hexanediamine (1,6-D) has been isolated in the period of germination of young radish seeds. Production of thiamine is higher during germination.

AGROTECHNIQUES

Radishes originated in the temperate zone but can be cultivated from there right through to the tropics, where they grow best at latitudes higher than 10° N and S, or at elevations above 500 metres. The plant grows best in areas where annual daytime temperatures are within the range 12 - 25°c, but can tolerate 3 - 30°c. It prefers a mean annual rainfall in the range 800 - 1,000mm, but tolerates 500 - 2,800mm.Very easily cultivated fast-growing plants which prefer a sunny position and a rich light soil with ample moisture. The root is best harvested before the plant flowers.

32. Common name: Jowan, caraway, ajowan
Scientific name: *Trachyspermum ammi*
Family: Apiaceae

Introduction

Ajowan is an erect, aromatic, branched annual herb growing 60 – 160 cm tall. The plant is often grown, especially in southern Asia, for its seed which is used as a spice. It also has a wide range of medicinal applications.

Habitat

It is cultivated in dry or semi dry areas mainly.

Medicinal Properties

The seed, and especially the essential oil in the seed, is strongly antiseptic, antispasmodic, aromatic, bitter, carminative, diaphoretic, digestive, diuretic, expectorant and tonic. It is used internally in the treatment of colds, coughs, influenza, asthma, diarrhoea, cholera, colic, indigestion, wind, oedema, arthritis and rheumatism. Thymol, present in essential oil, is strongly antiseptic and antibacterial against many gram-positive and gram-negative bacteria. It is also showing antifungal activity against a range of storage and soil-borne fungi.Thymol has molluscicidal, nematicidal and larvicidal effects. It has been shown to have vasodilatory and hypotensive activity, producing a dose-dependent reduction in blood pressure and heart rate. In addition, it has shown effective anti-oxidant activity.

Other Uses

The pungently aromatic fruits are about 2 cm long. The pungent flavour is said to be a combination of aniseed and oregano with a hint of black pepper. They are widely used, especially in India, the Mediterranean and Ethiopia, as flavouring in savoury dishes, including curries, pulses, breads and pastry snacks. The fruit is usually dried, then roasted and ground into a powder before being used as a spice.

BOTANY

Ajwain (*Trachyspermum ammi* L. Sprague) is an annual aromatic and herbaceous plant of the family Apiaceae. It is an erect annual herb with a striate stem and originated in the eastern regions of Persia and India. Its fruits are small, and grayish-brown in color.

CHEMICAL CONSTITUENT

The seed contains about 4 - 6% essential oil. The ajwain essential oil contains thymol (87.75%) and carvacrol (11.17%) as major constituents and major nonphenolic components quantified were p-cymene (60.78%) and γ-terpinene (22.26%).

AGROTECHNIQUES

Ajowan can be grown in the tropics and subtropics, where it is cultivated at elevations from near sea level up to 2,200 metres. It grows best in areas where annual daytime temperatures are within the range 16 - 24°C, but can tolerate 10 - 30°C. It prefers a mean annual rainfall in the range 800 - 1,000 mm. The crop requires a moist soil in a sunny position. It prefers loamy soils that are not too heavy, though it can be grown on all types of soils. Seeds usually takes 10 - 15 days germinate, though it may take as long as 1 month. Cool and cloudy weather and gentle rain after sowing are important for the establishment of the crop. The seed is often broadcast, as a sole crop at a rate of 2.3 - 3.5 kg per hectare, or intercropped with grain crops or other Apiaceae. The light fruits need to be covered by a thin layer of soil for good germination, and to prevent flushing. For sowing, the fruits are sometimes mixed with rice husks to ensure even distribution. Good seed is difficult to obtain, as many fruits for sale on the market are empty.

33. Common name: Salinche, Sessile Joyweed, Dwarf copperleaf
Scientific name: *Altenanthera sessilis*
Family: Amaranthaceae

INTRODUCTION

Alternanthera sessilis is an erect, ascending or creeping herbaceous plant that is usually perennial, though sometimes annual, growing up to 1 m tall. The plant is often widely branched, and has a vigorous taproot. The plant is often harvested from the wild for local use as a food and medicine.

HABITAT

A. sessilis typically grows in disturbed wetland habitats, riparian zones, estuarine habitats, open fields and plantations. The species can also grow in dry conditions and can grow along roadsides, pathways and fallow ground. In Africa and Asia, the species has been recorded as growing within fields of a number of cultivated crops (e.g., rice, maize, cotton).

Medicinal Properties

The plant is said to be abortifacient, cholagogue, febrifuge and galactagogue. It is eaten by nursing mothers who wish to increase their milk flow. An infusion of the entire plant is used as a remedy against intestinal cramps, fever, diarrhoea and dysentery. The juice of the plant is used to treat white discharge in the urine. The plant is used in mixtures with other medicinal plants, to treat hepatitis, tight chest, bronchitis, asthma, and lung troubles, to stop bleeding and as a hair tonic. It is also applied externally in the treatment of scabies, cuts and wounds, boils; and as a cooling agent to treat fever.

A paste of the plant is applied as a poultice on wounds, to draw out spines and other objects from the body. Mixed with corn flour and baked, it is eaten to treat menstrual disorders. The juice of the root is used in the treatment of dysuria, fevers and bloody dysentery. The plant is said to be abortifacient, cholagogue, febrifuge and galactagogue. It is eaten by nursing mothers favouring to increase their milk flow. An infusion of the entire plant is used as a remedy against intestinal cramps, fever, diarrhoea and dysentery. The juice of the plant is used to treat white discharge in the urine. The plant is used in mixtures with other medicinal plants, to treat hepatitis, tight chest, bronchitis, asthma, and lung troubles, to stop bleeding and as a hair tonic. It is also applied externally in the treatment of scabies, cuts and wounds, boils; and as a cooling agent to treat fever. In India and Sri Lanka, it is used for treatment of gastrointestinal problems. *A. sessilis* is used as a treatment for headaches and vertigo in Nigeria. It is also used to treat hepatitis, bronchitis, and asthma in Taiwan.

OTHER USES

Leaves and tender tops are consumed as raw (in salads, soups) or cooked (as potherb). It resembles spinach in flavour and appearance. The antioxidant carotene is found in large amounts in the plant. In Southeast Asia young shoots and leaves are ingested as vegetables. *A. sessilis* is a plant that is puchased for a water garden and over the internet for aquariums. Also, it is bought for its low cost and taste by the people of Sri Lanka. *A. sessilis* is chopped up when it is fed to ducks and is also fed to pigs.

BOTANY

This is a perennial herb with prostrate stems, 10 to 100 cm long, rarely ascending, often rooting at the nodes. Leaves are obovate to broadly elliptic, occasionally linear-lanceolate, 1–15 cm long, 0.3–3 cm wide, glabrous to sparsely villous, petioles 1–5 mm long. Flowers in sessile spikes, 0.7-1.5 mm long. *A. sessilis* has a somewhat glistening fruit, that is light-beige yellow. Fruits are utricles 1.8–3 mm long and 1.3–2 mm wide. Seeds are lenticular 0.9–1.5 mm long and 0.8–1 mm wide.

CHEMICAL CONSTITUENTS

The major components of essential oil of leaves are 1,1,1,5,5,5-hexamethyl-3,3-bis[trimethylsilyl) oxy]trisiloxane (15.43%), S,S-dioxide trans-2-methyl-4-N-pentylthiane (11.27%), didodecylphthalate (10.62%) and tetrahydro-2,5-dimethoxy furan (10.01%). However, the major components of essential oil of flower were 1,1,1,5,5,5-hexamethyl-3,3-bis[trimethylsilyl)oxy]trisiloxane (17.76%), trans-4-ethyl-5-octyl-2,2-bis(trifluromethyl)-1,3-dioxolane (11.12%) and tetrahydro-2,5-dimethoxy furan (9.10%).

AGROTECHNIQUES

The plant succeeds in sun or moderate shade. Prefers a loamy, alkaline soil, but can succeed in a wide range of soils. Plants can be grown in moist to wet soil and in shallow water. *Alternanthera sessilis* is a weed that inhabits many areas of the world. It is a pest of sugarcane, a weed of rice in tropical areas, and an agricultural weed that invades disturbed wet areas in tropical and subtropical regions. It is a rather variable plant, growing upright in drier soils, but becoming prostrate in wetter conditions with the shoots rooting at the nodes.

34. Common name: Sushni, European water clover
Scientific name: *Marsilea quadrifolia*
Family: Marsileaceae

INTRODUCTION

Water clover is a deciduous, aquatic, small, creeping fern producing clumps of leaves up to 15 – 20 cm tall with erect 4-foliolate leaves, arising solitary from the nodes on a long creeping rhizome, rooted at the nodes. It is usually planted as an ornamental pot plant and is used for pond decoration. *Marsilea minuta* has a wide native range. It is the most widely distributed and common fern in India with few or no threats. The plant forms dense stands, or has a random growing pattern. This species is widespread and common in Europe and some Asiatic countries such as China.

HABITAT

Lacustrine (in lakes or ponds), riverine (in rivers or streams), shores of rivers or lakes are the normal habitat. Originally from Europe, the plant has escaped from cultivation and spread to many areas of the tropics and subtropics. It has been labelled as 'Invasive' in many countries, and has become a major weed of paddy fields and a troublemaker in ponds, lakes and irrigation systems. It frequents paddy crops and warm, shallow, muddy edges of fresh water ponds.

MEDICINAL PROPERTIES

The juice made from the plant is anti-inflammatory, diuretic, depurative, febrifuge and refrigerant. The plant is also applied externally in the treatment of snakebites and skin injuries.

OTHER USES

Young stems and leaves may be used as famine food used in times of scarcity. The spores are produced in a sporocarp and in allied species this is ground up and mixed with flour to prepare bread etc. It is rich in starch. The sporocarps are also a rich source of thiaminase enzyme which can destroy vitamin B1 in the diet.

Botany

This herbaceous, perennial fern grows in aquatic environments from creeping, fleshing roots containing rhizomes. It has thin green stalk-bearing downy furled leaflets which will open to reveal four wedge-shaped leaflets. These 1.5 - 4 cm leaflets will float along the water surface.

Chemical Constituents

M. quadrifolia revealed the presence of tannins, saponins, flavonoids, steroids, terpenoids, alkaloids, carbohydrates, proteins and phenolic compounds. The other contents of phytocompounds include total phenolics, flavonoids, alkaloids, tannins and saponins in leaf and stem.

Agrotechniques

A plant originally of the temperate zone, it can be grown in many areas of the tropics and subtropics at elevations up to 1,500 metres. It grows best in areas where annual daytime temperatures are within the range 20 - 28°C. When dormant, the plant can survive temperatures down to about -15°C. It prefers a mean annual rainfall in the range 700 - 2,000 mm. It prefers full sun but is also very tolerant of shade. It requires a permanently moist or wet soil with a pH in the range 6.5 - 7.5. Mature plants bearing sporocarps can develop in as little as 3 months. Spore germination in the family occurs after rupture of the sporocarp wall allows the sporocarp contents to be hydrated. A gelatinous structure emerges from the sporocarp, breaking it into valves and carrying the sori into the water. Spore germination and fertilization occur immediately after division. Planting is usually done in May-June month. Application of nitrogen and phosphate can increase the yield.

35. Common name: Sufola, sathakuppai, sowa
Scientific name: *Anethum graveolens* L
Family: Apiaceae

Introduction

It is an erect, branched, annual herb growing up to 150 cm tall but more likely to be around 75cm. It is a commonly cultivated herb, especially in warm temperate and tropical zones. It is grown mainly for its edible leaves and seeds, though it is also used medicinally.

Habitat

Fields, waste places etc in the Mediterranean.

Medicinal Properties

The seeds are a common and very useful household remedy for a wide range of digestive problems. An infusion is more than ever effective in treating gripe in babies and flatulence in young children. The seed is aromatic, carminative, mildly diuretic, galactagogue, stimulant and stomachic. The essential oil of the seed relieves intestinal spasms and griping. Chewing the seed improves bad breath. It is also a useful addition to cough, cold and flu remedies. It can be used with antispasmodics such as *Viburnum opulus* to relieve period pains. It can also help to increase the flow of milk in nursing mothers.

OTHER USES

Leaves are consumed raw or cooked, as flavouring in salads, etc. The leaves can be harvested at any time the plant is growing, but are best just before the plant flowers. It can also be sprouted and used in breads, soups and salad dressings. The seed contains up to 4% essential oils. It is used in perfuming soaps, medicines and as a food flavouring. Some compounds of dill (d-carvone is mentioned as one of them), when added to insecticides, have greatly increased the effectiveness of the insecticides.

BOTANY

Anethum graveolens L. is the sole species of the genus Anethum, though classified by some botanists in the related genus Peucedanum as *Peucedanum graveolens* (L.). A variant called East Indian dill or *Sowa* (*Anethum graveoeloens* var *sowa* Roxb. ex, Flem.) occurs in India and is cultivated for its foliage as a cold weather crop throughout the Indian sub-continent, Malaysian archipelago and Japan.

CHEMICAL CONSTITUENTS

The plant contains 253 calories, 7.2 g water, 20 g protein, 4.4 g fat, 55.8 g carbohydrate, 11.9 g fibre, 12.6 g ash, 1784 mg calcium, 543 mg phosphorus, 48.8 mg iron, 451 mg magnesium, 208 mg sodium, 3308 mg potassium, 3.3 mg zinc, 0.42 mg thiamine, 0.28 mg riboflavin, 2.8 mg niacin and 1.5 mg vitamin B6 per 100g. Many compounds viz., α-phellandrene (38.85%), p-cymene (14.72%), γ-terpinene (13.96%), α-pinene (9.61%), limonene (6.45%), germacrene D (6.01%), dill ether (3.38%), β-myrcene (2.51%), and α-thujene (2.06%) are the main constituents of essential oil.

AGROTECHNIQUES

It is primarily a summer crop of temperate climates. In the tropics it grows at elevations from 200 - 2,000 metres. It is generally considered a long-day plant in terms of flower initiation, with a critical period of 11 - 14 hours. The annual rainfall or water requirement ranges from 500 - 1,700 mm. The plant does not tolerate frost, thriving when the monthly average temperatures is around 16 - 18°C. The minimum temperature for growth is about 7°C.

Sandy loam soil is better for its cultivation. It can tolerate a pH in the range 5.3 to 7.8. Cultivation is done two times in a year. Seed sowing is done in Sep-Oct and April-May. The crop is very much intolerant of root disturbance and should not be transplanted because it will then quickly run to seed. A regular supply of leaves can be obtained if successional sowings are made throughout the year. There are many named varieties.' Bouquet' is an American cultivar that has a prolific production of seeds. The sub-species *A. graveolens* sowa from India has a slightly different flavour. The flowers are very attractive to bees. After harvesting matured seeds are collected by threshing.

36. Common name: Sajina, horseradish tree
Scientific name: *Moringa oleifera* L
Family: Moringaceae

Introduction

Moringa (*Moringa oleifera* Lam.) is a multipurpose tropical tree. It is mainly used for food and has numerous industrial, medicinal and agricultural uses, including animal feeding. Fruiting occurs once in a year. Important varieties of moringa are Jafna, Kodikal, Mulanur etc. It can be cultivated by sowing good quality seed. Normally it can be cultivated by stem cutting by using stem cutting. Another species of moringa is Nojna. Its fruiting habit is throughout the year. Leaf production is also high. Medium or large type of fruits are found.

Habitat

In its native range, *M. oleifera* is found in the tropical moist deciduous forests, southern tropical dry deciduous forests and primary seral type forests in northern India. In Africa, it is often found growing along stream banks or in savannas. It also occurs along roadsides, in urban areas, wastelands and coastal areas.

Medicinal Properties

The horseradish tree is a nutritious, diuretic, laxative herb that is expectorant, increases milk flow, controls bacterial infections and is rubefacient when applied topically. It contains a potent antibiotic. The root is used as a vesicant. The alkaloid spirachin (a nerve paralyzer) has been found in the roots. The root juice is used internally in the treatment of asthma, gout, rheumatism, enlarged spleen and liver, bladder and kidney stones, inflammatory conditions. Externally, the root is used to treat boils, ulcers, glandular swellings, infected wounds, skin diseases, dental infections, snake bites and gout. The roots and bark are worn for cardiac and circulatory problems, as a tonic and for inflammation. The bark is an appetizer and digestive. The gum is demulcent, diuretic, astringent and abortifacient. It is used in cough syrups and in the treatment of asthma. The bark and gum are used in the treatment of tuberculosis and septicaemia. Flowers and immature fruits are said to be a good rubefacient. A decoction of the flowers is used as a cold remedy. The seeds are effective against skin-infecting bacteria *Staphylococcus aureus* and *Pseudomonas aeruginosa*. A number of compounds with medicinal properties have been isolated.

Other Uses

Young leaves, shoots and flowers, having a mustard-like flavour, are consumed as raw (added to salads) or cooked (as a potherb in soups and curries). The leaves are very nutritious, containing 7 - 10% protein and rich in vitamins, minerals and the sulphur-containing amino acids, methionine and cystine. The long (15 – 45 cm), bean-like pods are used in soups and curries, or made into pickles. The immature sweet flavoured seeds are eaten like peas. Mature seeds, when roasted or fried, are said to resemble peanuts in flavour. Oil obtained from the seeds is used in salads and cooking. The oil is clear and odourless and does not become rancid quickly. The pungent root is used like horseradish (*Armoracea rusticana*) as a hot flavouring in foods. The condiment when used in excess may be harmful. A reddish gum obtained from the bark is used as a seasoning.

BOTANY

Moringa is a slender softwood tree that branches freely and can be extremely fast growing. Although it can reach 3 heights in excess of 10 m and a diameter of 20-40 cm at chest height, it is generally considered a small- to medium-size tree.

CHEMICAL CONSTITUENTS

It contains oxalic acid (fruit and leaf), moringinine (bark), vanillin (stem), kaempferol and quercetin (flower) and spirochin and pterygospermin (root). The seeds contain a glucosinolate that on hydrolysis yields 4-(alpha-L-rhamnosyloxy)-benzylisothiocyanate, an active bactericide and fungicide. It contains9-octadecenoic acid (20.89%), L-(+)-ascorbic acid- 2,6-dihexadecanoate(19.66%), 14 – methyl -8-Hexadecenal (8.11%) , 4- hydroxyl-4-methyl-2-pentanone (7.01%), 3- ethyl-2, 4-dimethyl-pentane (6.14%), phytol (4.25%), octadeamethyl-cyclononasiloxane (1.23%), 1, 2-benzenedicarboxylic acid (2.46%), 3, 4-epoxy-ethanone comprising (1.78%), N-(-1-methylethyllidene)-benzene ethanamine (1.54%), 4, 8, 12, 16- tetramethylheptadecan-4-olide (2.77%), 3-5- bis (1, 1-dimethylethyl)-phenol (2.55%), 1- hexadecanol (1.23%), 3, 7, 11, 15- tetramethyl-2 hexadecene-1ol (1.17%), hexadecanoic acid (2.03%) and 1, 2, 3- propanetriyl ester-9 octadecenoic acid (1.23%) as the chemical composition in the leaf extract.Also oleic acid (84%) L-(+)-ascorbic acid- 2, 6-dihexadecanoate (9.80%), 9-octadecenoic acid (1.88%), methyl ester hexadecanoic acid (1.31%) and 9- octadecenamide (0.78) are found in the seed extract.

AGROTECHNIQUES

Horseradish tree grows well in warm temperate to tropical areas and can be found at elevations from sea-level to about 1,000 metres. It comes up best in areas where annual daytime temperatures are within the range 20 - 35°C, but can tolerate extremes upto 7 - 48°C. The plant is quite cold hardy and is not harmed by light frosts, but it can be killed back to ground level by a freeze. It prefers a mean annual rainfall in the range 700 - 2,200 mm. It can be easily grown in a well-drained soil in a sunny position and can tolerate a wide range of soil types. It is in general suitable for light, medium and heavy soils, but cannot withstand salinity. It prefers a pH in the range 5.5 - 7. Established plants spaced at 2.4-3 m are quite drought tolerant but yield much less foliage when continuously under water stress. It grows better if given shelter from strong winds. Moringa is an extremely fast-growing tree. Within 1- 3 months, trees can reach 2.5 metres in height. A growth rate of 3 - 4 metres per year is not unusual for young plants. Young trees raised from seed start flowering after 2 years. In trees grown from cuttings the first fruits may be expected 6 - 12 months after planting. Flowering often precedes or coincides with the formation of new leaves, and can occur throughout the year in non-seasonal climates. Constant pruning of up to 1.5 metres per year is suggested to obtain a thick-limbed and multibranched shrub. It coppices and pollards well. The sweet smelling flowers are produced throughout the year. In the warmer parts of its range the plant can produce a second crop of seeds each year.

37. Common name: Setberale sak, arrowleaf sida, jelly leaf
Scientific name: *Sida rhombifolia* L
Family: Malvaceae

INTRODUCTION

Broom jute is a perennial plant producing more or less herbaceous, many-branched stems around 1 m long from a strong, woody rootstock. The stems can be erect or prostrate and are usually more or less woody. The plant is gathered from the wild as a local source of medicines and fibre.

HABITAT

A common and widespread weed of disturbed sites, roadsides, pastures, grasslands, open woodlands, crops, waste areas, footpaths and gardens in tropical, sub-tropical, warmer temperate and semi-arid environments.

MEDICINAL PROPERTIES

A decoction of the whole plant is used as a treatment for fevers. A paste of the plant is used to treat indigestion and also used as a poultice in the treatment of headaches, boils, cramps, rheumatism, toothache, chapped lips and pimples. The plant is ground and mixed with soft grease and sugar to make a poultice that is applied to soften abscesses and release paste leaves are diuretic. An infusion is used to treat dysentery. The juice of the leaves is mixed with vinegar to make an anti-inflammatory and digestive remedy. A decoction of the leaves is used to wash wounds. The leaves are applied to the head as a poultice to remedy headache. The yellow flowers are eaten with wild ginger in order to ease labour.The root is scraped into sea water and the mixture drunk as a treatment for diarrhoea, dysentery and abdominal upsets. A paste of the root is applied to boils.

OTHER USES

Broom jute is a perennial plant producing more or less herbaceous, many-branched stems around 1 m long from a strong, woody rootstock. The stems can be erect or prostrate and are usually more or less woody. The plant is gathered from the wild as a local source of medicines and fibre. In Australia, it is one of the best recognized native remedies for diarrhoea.

BOTANY

The stems are erect to sprawling and branched, growing 50 to 120 centimeters in height, with the lower sections being woody. The dark green, diamond-shaped leaves are arranged alternately along the stem, 4 to 8 centimeters long, with petioles that are less than a third of the length of the leaves. The leaves are paler below, with short, grayish hairs. The apical halves of the leaves have toothed or serrated margins while the remainder of the leaves are entire (untoothed). The petioles have small spiny stipules at their bases.

Chemical Constituents

The plant contains cryptolepine, ephedrine and vasicine. The leaves contain around 7.4% protein. A good quality fibre obtained from the bark is used for making ropes and twine.

Agrotechniques

It grows wild in a range of soil types, from fertile to degraded condition. The awned seeds are spread by adhering to clothing and livestock, in mud on vehicles, and as contaminants in hay and seed crops. April-May months are ideal for its cultivation. Before seed sowing ploughing of land is needed. The stems are gathered in the morning, tied into bundles and then used as brooms.

38. Common name: Helencha, buffalo spinach
Scientific name: *Enhydra fluctuans*
Family: Asteraceae

Introduction

Enydra fluctuans is a perennial herb with 30 – 80 cm height. The plant is harvested from the wild for its edible leaves, and is also popularly cultivated in Java, Malaya Cambodia and other countries.

Habitat

Swamps and aquatic areas in coastal regions of eastern Australia and marshes and streamsides in southern China are the natural habitats.

Medicinal Properties

Antidiabetic Uses and Antimicrobial Activity

Enhydra fluctuans extract can be effectively used as an antidiabetic plant by boiling and cutting at the nodes. The tribal practitioners of the Marakh sect of the Garo tribe living in Mymensingh of Bangladesh uses twelve medicinal plants for treatment of diabetes out of which *Enhydra fluctuans* is one. It has been found to have potent antibacterial activity against many gram positive as well as gram negative organisms. Some of them are *Escherichia coli*, *Pseudomonas aeruginosa*, *Staphylococcus typhi*, etc.

Anti-Inflammatory Activity

The flavonoid isolated from leaves of *Enhydra fluctuans* shows anti inflammatory activity by inhibiting COX-2 and 5-LOX. Moreover, these flavonoids exhibit *in vitro* on key enzymes of arachidonic acid cascade involved in the mediation of inflammation. Based on it, *in vivo* anti-inflammatory activity of flavonoid fraction was evaluated by using carrageenan induced paw oedema and cotton pellet induced granuloma. It can reduce the inflammation significantly.

Anti-Oxidant Activity

The antioxidant potential of crude methanol extract, ethyl acetate and n-butanol soluble fractions of *Enhydra fluctuans*, is widely used. The crude extract and all the fractions showed free radical scavenging activity when tested in different models. Among all the fractions, the ethyl acetate fractions exhibited the highest free radical scavenging activity in all the tested models comparing to crude extract and n-butanol fraction.

Anti-Cancer Activity

Flavonoids obtained from *Enhydra fluctuans* (FEF) were screened for anticancer activity against Ehrlich's ascites carcinoma (EAC) bearing Swiss albino mice. Two flavonoids, baicalein 7-O-glucoside and baicalein 7-O-diglucoside, were isolated from the ethyl acetate fraction. Treatment with FEF caused a significant decrease in the tumour cell volume and increase of life span.

Anti-Diarrhoeal Activity

The methanolic and aqueous extract was evaluated in experimental diarrhoea, induced by castor oil in mice. Both methanolic and aqueous extracts, given orally at a dose of 250 mg/kg body weight, showed significant anti-diarrhoeal activity.

OTHER USES

The young parts are used as a salad in several countries, including Malaya. Sometimes they are steamed before they are eaten. In the Philippines, the leaves are pressed and applied to the skin as a cure for certain herpetic eruptions. In West Bengal it is washed, chopped and cooked as Sag fry or boiled with rice and eaten with boiled rice with boiled potato, salt and mastered oil. The young parts and the leaves of the plant are somewhat bitter and are used by the Malays as a laxative. The leaves are useful in diseases of the skin and of the nervous system. The fresh juice of the leaves is prescribed as an adjunct to tonic metallic medicines, and is given in neuralgia and other nervous diseases. The leaves are antibilious. The expressed juice of the leaves is used as a demulcent in cases of gonorrhea. As a cooling agent, the leaves are pounded and made into a paste which is applied cold to the head.

BOTANY

It is a perennial, often gregarious, erect-prostrate or free-floating, aromatic herb having 30-100 cm height. Stem is cylindrical, 0.5-1 cm in diameter, hollow, hairy, sparsely branched, rooting at the nodes. Leaves opposite, (sub-) sessile, narrowly oblong, 2-10 cm × 0.5-2 cm, base broadly truncate or subcordate, apex narrowly obtuse, margins usually dentate-serrate, finely gland-dotted on both surfaces. Inflorescence a globular head, up to 1 cm in diameter, sessile, solitary, terminal or pseudo-axillary (due to the development of axillary branches), heterogamous, many-flowered; involucre of 4 foliaceous, longitudinally-veined bracts, longer than the flowering heads and completely enclosing them. Ray flowers in many whorls, female; corolla with long tube and 2-5-lobed limb, white or greenish-white; ovary covered by palea which is ciliate at top, style ultimately exserted. Disk flowers more or less numerous, bisexual or sometimes sterile; corolla campanulate with 5-lobed rim; ovary as in ray flower, style arms 2; stamens connate, anthers ultimately exserted. Fruit an achene, oblongoid, 3 mm long, glabrous, closely clasped by the thickened palea (acting as floater), blackish, without a pappus.

CHEMICAL CONSTITUENTS

Different extracts of *Enhydra fluctuans* have been tested for the presence or absence of primary and secondary bioactive compounds like carbohydrates, proteins, oils, alkaloids, flavonoids to name a few. It has been found to be a rich source of flavonoids and moderate presence of alkaloids, tannins, phenolics and carbohydrates have been reported. The plant is also a rich source of β-carotene and was also found to be high on protein content. Presence of saponin too has been stated. A novel bioactive isoflavone glycoside 4',5,6,7-tetrahydroxy-8- methoxyisoflavone-7-O-β-D galactopyranosyl-(1→3)-O-β-Dxylopyranosyl-(1→4)-O-α-l-rhamnopyranoside has been reported from the methanolic extract of the leaves of *Enhydra fluctuans* Lour. Also the aqueous extract made known the stimulation of a biological response of the carbohydrate polymer of the plant. Based on spectral and chemical evidence the structure of Enhydrin, a germacranolide has been isolated from the leaves and stems of *Enhydra fluctuans*. The structure and stereochemistry of sesquiterpene lactones were elucidated by NMR techniques and chemical reactions.

AGROTECHNIQUES

Little is known about growth and development. In Java it is reported to flower year-round, in Vietnam from June to September. The fruits often do not develop.

39. Common name: Kalkasunda, Coffeeweed, Negro coffee
Scientific name: *Cassia occidentalis L*
Family: Fabaceae

INTRODUCTION

It is a very leafy, scented, shrubby plant growing up to 2 metres. The plant is often annual at least in seasonal climates. It is sometimes collected from the wild basically for local medicinal use. The seed is sometimes used as a coffee substitute.

HABITAT

A weed of roadsides, waste areas, disturbed sites, pastures, grasslands, open woodlands, coastal environs and crops in tropical, subtropical and semi-arid regions.

MEDICINAL PROPERTIES

The whole plant is diuretic, febrifuge, stomachic and tonic. It is used in the treatment of hypertension, dropsy, diabetes, fevers, biliousness, rheumatism, ringworm and eczema. The plant is boiled and applied externally, pounded and mixed with wood-ash, and rubbed on areas of leishmaniasis and eczema. The leaves, seed, root and skin can be used as medicine. In Ayurveda, the juice is regarded to be digestive. It is gargled for treating throat troubles. The root is cholagogue, emetic and purgative. An infusion is used in the treatment of bilious fever, ordinary fever, stomach-ache, and to ease menstruation. A tincture of the root is rubbed onto rheumatic areas. A tea made from roots and dried flowers is used as a treatment for colds and upset stomach. The leaf is used as a remedy for renal calculi. Leaves are made into a tea for treating afterbirth problems, fevers, coughs and colds, headaches, haemorrhage and thrush. An ointment prepared from the leaves is applied as a remedy for ringworm and other affections of the skin. The flowers are used in a preparation to reduce stomach acid in children. The seed is febrifuge and sedative. An infusion is drunk to calm the nerves, and as a treatment for kidney problems, haemorrhage, worms, and cleaning womb and tubes. Extracts of the leaf and seed have shown antibiotic activity.

Other Uses

The seeds are dried, roasted then ground into a powder and used as a coffee substitute. It is called Mogdad or Negro coffee. Young leaves - steamed and eaten as a vegetable. Immature seedpods and flowers can be steamed and eaten as a vegetable.

Botany

Leaves are evergreen, lanceolate, compound, glossy leaf surface, deep tap root system, stem is hard and woody. The plant gives a special smell by rubbing leaves. Flowers are yellow in colour with 1 to 2 cm diameter arranged in raceme type inflorescence, axillary and also forming terminal, bracts are caduceus. Fruits are flat pods, 10-12 cm long with 10-30 seeds.

Dicot seed type with characteristic odour and bitter taste. Areolate seeds are pointed at end and blunt at the other.

Chemical Constituents

The main plant chemicals in *C. occidentalis* include: achrosin, aloe-emodin, emodin, anthraquinones, anthrones, apigenin, aurantiobtusin, campesterol,hrysophanol, chrysoeriol, emodin, essential oils, funiculosin, galactopyranosyl, helminthosporine, islandicine, kaempferol, lignoceric acid, linoleic acid, linolenic acid, mannitol, mannopyranosyl, matteucinol, obtusifolin, obtusin, oleic acid, physcion, quercetin, rhamnosides, rhein, rubrofusarin, sitosterols, tannins, and xanthorine. The leaf contains flavonoid glycosides, an anthraquinone, and a bianthraquinone.The seed contains N-methyl morpholine, campesterol and beta-sitosterol glucosides. For example stems, leaves and root bark of the plant from Ivory Coast, Africa contain small amount of saponins, no alkaloids, sterols, triterpenes, quinines, tannins and flavonoids. However, a large amount of alkaloids were found in the stem, leaves and fruits from Ethiopia.

Agrotechniques

It has a wide native range from the subtropics of S. America through to southern N. America and the Caribbean. It grows primarily at lower altitudes. It favours moister areas in the tropics. It grows in areas where the mean annual rainfall is within the range 1,000 - 2,200 mm. It essentially prefers a deep, well-drained, moderately fertile sandy loam and a position in full sun. Originally native of the Americas, the plant is widely cultivated through the tropics and subtropics and is commonly naturalized in many areas. While coffee senna can be a weed of cultivated fields and plantations, it is principally a problem because it accumulates in heavily grazed pastures. The plant flowers throughout the year in non-seasonal climates. The plant has a disagreeable odour when crushed.

40. Common name: Cress, pepperweed
Scientific name: *Lepidium sativum*
Family: Brassicaceae

INTRODUCTION

Garden cress is a profusely-branched, erect, annual plant growing up to 80cm tall. The plant has been cultivated for food since before the ancient Egyptians and is still widely used. It is most commonly eaten in the seedling form, within 7 days of sowing the seed. It is grown in Ethiopia for the edible oil obtained from its seed.

HABITAT

Rubbish tips, wasteland, roadsides, bird-feeding places, yards, gardens.

MEDICINAL PROPERTIES

The seeds are aperient, aphrodisiac, diuretic, emmenagogue, galactogogue, laxative and tonic. They have been boiled with milk and used to procure an abortion. A paste of the seed is taken, with honey, to treat amoebic dysentery and diarrhoea. The seeds are chewed to treat sore throat, cough, asthma and headache, and in large quantities to induce an abortion. Crushed seeds in water are used to treat hiccough and stomachache, for pains and hurts (as poultice) and for sunburn and other skin problems (as paste). The pounded seeds are poulticed on the skin, and have a vesicant and soothing action on bruises and sprains. The leaves are antiscorbutic, diuretic and stimulant. They are useful in the treatment of liver complaints. The plant is administered in cases of asthma, cough with expectoration and bleeding piles. The root is used in the treatment of secondary syphilis and tenesmus. The ethanol extract of the seeds showed significant antiinflammatory effects. The seed oil has a pronounced oestrogenic activity. The antibacterial action of *Lepidium sativum* has been demonstrated in several tests which depend largely on the age of the plant. The extract of the fresh leaves have strong antibacterial action against *Bacillus subtilis* and *Micrococcus pyogenes*, but is less effective against *Escherichia coli*.

OTHER USES

Young leaves are eaten as raw (salad) or cooked. They take about 7 days to be ready and have a pleasantly hot flavour. The boiled seeds are consumed in drinks by Arabs, either ground in honey or as an infusion in hot milk. The root is used as a condiment. The fresh or dried seedpods can be used as a pungent seasoning. The edible oil obtained from the seed has antioxidant properties. The seed yields up to 58% of edible oil that can also be used for lighting and making soap. The seeds are applied externally as an insect repellent. *Lepidium sativum* is extensively used as a test organism in plant physiological studies, as an indicator organism to examine toxicity levels of environmental pollutants, and in experimental studies assessing diverse pathogens.

BOTANY

Garden cress is an annual erect herbaceous plant growing up to 50 cm. The basal leaves have long petioles and one lyreate pinnatipartite; the coulinar leaves are lanceonate. A degree of variability is noted in the character of the basal leaves, which are cleft or split to a greater or lesser degree, a character, which is controlled by a single incompletely dominant gene. The inflorescence is in dense racemes. The flowers have white or slightly pink petals, measuring 2 mm. The siliquae measure 5 to 6 mm, are elliptical elates from the upper half and glabrous. It is an allogamous plant with self-compatible and self-incompatible form and cut the various degrees of tolerance to prolonged autogamy. There are diploid ($2n = 2x = 16$) and tetraploid ($2n = 4x=32$) forms.

CHEMICAL CONSTITUENTS

The leaf contains proteins, fat, carbohydrates, minerals – calcium and phosphorus, trace elements such as iron, nickel, cobalt and iodine, also contains various vitamins such as vitamin A, thiamine, riboflavin, niacin and ascorbic acid. The aerial parts of the plant contain stigmast-5-en-3β, 2-7-diol- 2-7-benzoate as one of the key chemical constituent. The plant also contains glucotropoeolin, 4-methoxyglucobrassicin, sinapine, sinapic acid, calmodulin, sinapoyglucose, esters of caffeic, pcoumaric, ferulic, quinic acids, protein, minerals, vitamins, 5-4'-dihydroxy-7,8,3',5-tetramethoxyflavone, 5-3'-dihydroxy-7,8,4'- trimethoxyflavone, 5-3'-dihydroxy-6,7,4'- trimethoxyflavone. The stem and leaves contain glucosinolates, the main component being glucotropaeolin (benzyl glucosinolate). Upon steam distillation the herb yields about 0.1% of a colourless essential oil, with a characteristic pungent odour. Seeds of the plant mainly contains alkaloids viz., lepidine, glucotropaeolin, N,N'-dibenzyl urea, N, N'-dibenzylthiourea, sinapic acid and its choline ester (sinapin). They also contain carotene, cellulose, calcium, phosphorus, iron, thiamine, riboflavin, niacin, uric acid. The seed coat of germinating seeds is rich in mucilage, which contains an allelopathic substance, lepidimoide. Seed oil contains palmitic, stearic, oleic, linoleic, arachidic, behenic, lignoceric acids, benzyl isothiocyanate, benzyl cyanide, sterol and sitosterol. The oil is rich in oleic, linoleic and uric acids, and also contains imidazole alkaloids.

AGROTECHNIQUES

Garden cress is cultivated from the temperate zone (as a spring to autumn crop) to the tropics, where it succeeds all year round but generally grows best in the cool season or at higher elevations around 2,400 metres. Plants require some protection if temperatures fall below -5°C. An easily grown plant, it comes up in most soils but essentially prefers a moist soil and also some shade during the summer to prevent premature flowering. The life cycle of the plant is about 3 months. Garden cress is often cultivated as a sprouted seed. A very easy and fast crop, it can be ready within 7 days from sowing the seed. It can also be grown outdoors as full grown plants and can provide fresh leaves for the salad bowl all year round from successional sowings. In water, the seeds become covered in slime.

41. Common name: Leek
Scientific name: *Allium porrum*
Family: Amaryllidaceae

Introduction

Its onion like vegetable and has a mild, sweet onion like flavour. Cultivation is more concentrated in plain lands. Cultivation is done in hilly tropical areas also. The leek is an ancient crop and is native to eastern Mediterranean lands and the Middle East.

Habitat

Suitable for light (sandy) and medium (loamy) soils, prefers well-drained soil and can grow in heavy clay soil. Suitable pH ranges wide such as acidic, neutral and basic (alkaline) soils and can also grow in many alkaline soils. It cannot grow in the shade.

Medicinal Properties

It has a very long folk history against a wide range of ailments, such as ringworm, and also having fungicidal, antiseptic, tonic, parasiticidal and even anticancer properties. The bulb is said to be anthelmintic, antiasthmatic, anticholesterolemic, antiseptic, antispasmodic, cholagogue, diaphoretic, diuretic, expectorant, febrifuge, stimulant, stomachic, tonic and vasodilator. The crushed bulb may be applied as a poultice to ease the pain of bites, stings etc.The whole plant is said to repel insects and moles.

OTHER USES

The leaves and long white blanched stem are eaten cooked. They can also be cut into thin slices and be added to salads. Bulbs are consumed as raw or cooked but flowers are eaten raw only. It is used as a garnish on salads.

BOTANY

Allium porrum is a bulb growing to 0.9 m. The cultivated leek is a tetraploid (4x=32) whereas the onion is diploid (2n=16). Like garlic, the leaves are dorsiventrally flattened rather than being radial. In the plant's first season of growth, long linear leaves arise from a compressed stem plate; the thick leaf bases overlap and are arranged concentrically in a nearly cylindrical bulb. A tuft of fibrous shallow roots grows from the base of the stem. It left unharvested, second season leeks produce a large umbel with many flowers; the seeds are small, black, irregular and angular.

CHEMICAL CONSTITUENTS

The phytochemical screening of the plant showed the presence of tannin, saponins, flavonoids, quinine, glycoside, cardiac glycoside, terpenoids, phenols, coumarins, steroids, alkaloids, anthocyanin and betacyanin. Ethanol and acetone leaf extract exhibits highest positive response followed by other solvent such as chloroform, petroleum ether and aqueous extract.

AGROTECHNIQUES

The leek is a widely cultivated vegetable. Although commonly treated as a biennial, this plant is a true perennial, perennating by means of small lateral growths and often developing a roundish bulb at the base of the main growth. It prefers an open sunny place in a light well-drained soil. It can tolerate a pH in the range 5.2 to 8.3. Young plants are often planted quite deeply in the soil (8 – 10 cm) in order to blanch the lower stem.

42. Common name: Celery, Ajmod, Ajwain-ka-patta
Scientific name: *Apium graveolens*
Family: Apiaceae

INTRODUCTION

Celery is very popular all over India similar to coriander. The seeds have a characteristic aroma with pungent taste and are used as a condiment in the flavoring of food products. This is a very important spice in cold area.

HABITAT

The *Apium graveolens* is a species native to the Mediterranean area and known as a medicinal plant since the time of Homer. In Italy it is found cultivated throughout the territory but rarely subspontaneous. Its natural habitat is that of the vegetable gardens, ruins and uncultivated, from 0 to 1500 meters.

MEDICINAL PROPERTIES

Wild celery reduces blood pressure, relieves indigestion, stimulates the uterus and is anti-inflammatory. It is also used as an efficient antidote for a range of ailments such as bronchitis, liver and spleen disease, arthritic pain. The herb is believed to be useful against hysteria. The essential oil obtained from the plant has a relaxing effect on the central nervous system. Some of its constituents have antispasmodic, sedative, anticonvulsant and even anticarcinogenic actions. A homeopathic remedy is made from the herb. It is used in treating rheumatism and kidney complaints, high blood pressure, bronchitis, asthma, liver, and spleen diseases. The ripe seeds, herb and root are aperient, carminative, diuretic, emmenagogue, galactogogue, nervine, stimulant and tonic. It should never be prescribed for pregnant women.

OTHER USES

Celery is normally consumed as a vegetable across the world. The leaves are strongly flavored and are used as a flavoring in soups and stews or as a dried herb. In Europe the hypocotyl is used as a root vegetable. Celery, onions, and bell peppers are the "holy trinity" of Louisiana Creole and Cajun cuisine. Whereas, celery, onions and carrots comprise the French mirepoix, habitually used as a base for sauces and soups.

BOTANY

Celery is an aromatic herb. The leaves are precisely pinnate to bipinnate with rhombic leaflets having dimension 3–6 cm X 2–4 cm (length X breadth). The flowers, creamy-white and 2–3 mm in diameter, are produced in dense compound umbels,. The seeds are broad ovoid to globose, 1.5–2 mm long and wide. A celery stalk readily separates into "strings" which are bundles of angular collenchyma cells exterior to the vascular bundles.

CHEMICAL CONSTITUENTS

The major bioactive compounds in the celery include a class of phenolic compounds called furanocoumarins. The main linear furanocoumarins in celery include the three phototoxic furanocoumarins viz., psoralen, xanthotoxin and bergapten.

AGROTECHNIQUES

The saplings should be made by sowing seed. It should be sprout within 3-4 weeks. Organic & chemical manure are used. The field should be made well with 500gm saplings per bigha. The polythene sheet is used to discolour.

43. Common name: *Lettuce,* milkweed
Scientific name: *Lactuca sativa L*
Family: Asteraceae

Introduction

This hardens and dries when in contact with the air. The sap contains 'lactucarium', which

Medicinal Uses

It is used in medicine for its anodyne, antispasmodic, digestive, diuretic, hypnotic, narcotic and sedative properties. It is taken internally in the treatment of insomnia, anxiety, neuroses, hyperactivity in children, dry coughs, whooping cough, rheumatic pain etc. An infusion of the fresh or dried flowering plant can also be used. The plant should be used with caution, and never without the supervision of a skilled practitioner. Even normal doses can cause drowsiness whilst excess causes restlessness and overdoses can cause death through cardiac paralysis. The sap has also been applied externally in the treatment of warts. The seed is anodyne and galactogogue. Lettuce has acquired a folk reputation as an anaphrodisiac, anodyne, carminative, diuretic, emollient, febrifuge, hypoglycaemic, hypnotic, narcotic, parasiticide and sedative.

Other Uses

Leaves are consumed as raw or cooked. Seeds are sprouted and used in salads or sandwiches. Edible oil is obtained from the seed. The seed is very small; extraction of the oil on any scale would not be very feasible.

Botany

Lactuca sativa is further characterized by a yellow corolla, white pappus, involucral bracts erect in fruit, and a flat cypsela body that has 5- to 9-nerved faces, the body terminated by an elongate, filiform beak.

Over the centuries a number of more or less distinct forms have arisen in cultivation. These forms have been classified as follows;

- *L. sativa angustana.* L.H.Bailey. is the Celtuce. The leaves of this form are not of such good quality as the other lettuces and the plant is grown more for its thick central stem which is used in the same ways as celery.
- *L. sativa capitata.* L. is the heading lettuce, it forms a heart in a similar way to cabbages. Examples of this include the Iceberg and Butterhead lettuces.
- *L. sativa crispa.* L. is the curled or leaf lettuce. This does not form a central heart but produces a loose rosette of basal leaves. It can be harvested on a cut and come again basis.
- *L. sativa longifolia* Lam. is the cos lettuce. This has longer, thinner leaves and a more erect habit; it does not form a compact heart.

Chemical Constituents

The essential oils isolated from the fresh and dry leaves as α-pinene (5.11% and 4.05%), γ-cymene (2.07% and 1.92%), thymol (11.55% and

10.73%), durenol (52.00% and 49.79%), α-terpinene (1.66% and 1.34%), thymol acetate (0.99% and 0.67%), caryophyllene (2.11% and 1.98%), spathulenol (3.09% and 2.98%), camphene (4.11% and 3.65%), limonene (1.28% and 1.11%) representing these major chemical compounds. However, some other minor chemical constituents were also isolated and identified from the essential oil of lettuce including β-pinene, α-terpinolene, linalool, 4-terpineol, α-terpineol, o-methylthymol, L-alloaromadendrene and viridiflorene.

AGROTECHNIQUES

It prefers a light sandy loam. The plant succeeds in most well-drained, humus-rich soils but dislikes acid conditions. Early and late sowings are best in a sunny position, whereas summer crops are grows well with some shade in order to slow down the plants tendency to go to seed and to prevent the leaves becoming bitter. The garden lettuce is widely cultivated in many parts of the world for its edible leaves and is probably the most commonly grown salad plant. There are many named varieties capable of providing fresh leaves throughout the year if winter protection is given in temperate areas. In wet weather they are likely to develop fungal diseases. In addition, the seed needs to be sown at regular intervals of 2- 3 weeks during the growing season in order to provide a regular supply of leaves. Lettuces make a good companion plant for strawberries, carrots, radishes and onions. They also grow well with cucumbers, cabbages and beetroot.

Concentrations of lactucarium are low in young plants and most concentrated when the plant comes into flower. It is collected commercially by cutting the heads of the plants and scraping the juice into china vessels several times a day until the plant is exhausted. The cultivated lettuce does not contain as much lactucarium as the wild species, most being produced when the plant is in flower.

44. Common name: Gajor Sak, bird's nest, bishop's lace
Scientific name: *Daucus carota* L
Family: Apiaceae

Introduction

Wild carrot of the family Apiaceae also includes parsnip, parsley, fennel and angelica. Wild forms have thin, wiry taproots, bearing little resemblance to the bright orange, fleshy root vegetable available commercially, although both share the characteristic carrot fragrance. Delicate white flower heads are produced after the second year of growth. The orange colour of cultivated carrots is due to a high concentration of beta-carotene. This is a precursor of vitamin A, which is important for growth, development and good vision.

Habitat

Habitats include overgrown thickets, degraded prairies, weedy meadows, areas along railroads, grassy roadsides, lawns, pastures, abandoned fields, fence rows, vacant lots, junk yards, and other waste areas. Fire is not very effective in removing this plant from natural areas, however it tends to decline spontaneously in such areas when there is an absence of disturbance. Once dispersed to an area, the seedlings can emerge and survive in several types of ground cover, including those with thick vegetation. It is commonly found in fields 4-7 years after abandonment.

MEDICINAL PROPERTIES

The wild carrot is diuretic, soothes the digestive tract and stimulates the uterus. The spiny fruits are used in the treatment of kidney complaints and dropsy. It is a wonderful cleansing medicine. It can support the liver, stimulates the flow of urine and regulates the removal of waste by the kidneys. The whole plant is anthelmintic, carminative, deobstruent, diuretic, galactogogue, ophthalmic, stimulant. An infusion is used in the treatment of various complaints viz., digestive disorders, kidney and bladder diseases and dropsy. An infusion of the leaves has been used to counteract cystitis and kidney stone formation. Porphyrins present in the carrot leaves stimulate the pituitary gland and lead to the release of increased levels of sex hormones. A warm water infusion of the flowers has been used in the treatment of diabetes. The grated raw root is used as a remedy for threadworms. Delayed menstruation is encouraged by the root. The seeds are diuretic, carminative, emmenagogue and anthelmintic. An infusion is used in the treatment of oedema, flatulent indigestion and menstrual problems. Carrot seeds can be abortifacient and so should not be used by pregnant women.

OTHER USES

Daucus carota roots are eaten raw and cooked. The seeds are used as aromatic agents in food items. The dried roots are roasted and used to make coffee. It is also used for making salads and stews. Flowers are mixed with thick batter and deep fried as fritters. Fresh flowers are also used for making jelly and seeds are used as toppings on bread. *Daucus carota* is used as a flavouring agent in many food items such as baked food, meat and meat products, etc. Roots are also used to make coffee. A solution made from seeds and roots is for making an insecticide. The essential oil derived from this plant is used to make anti-wrinkle creams. This plant is also cultivated for ornamental purposes.

BOTANY

Carrot (*Daucus carota*) is a herbaceous, usually biennial plant of the Apiaceae family that produces an edible taproot. Among common varieties, root shapes range from globular to long, with lower ends blunt to pointed. Besides the orange-coloured roots, white-, yellow-, and purple-fleshed varieties are known.

CHEMICAL CONSTITUENTS

The moisture content of carrot varies from 86 to 89%. Carrots are a good source of carbohydrates and minerals like Ca, p, Fe and Mg. The chemical constituents of carrot as moisture (86%), protein (0.9%), fat (0.2%), carbohydrate (10.6%), crude fiber (1.2%), total ash (1.1%), Ca (80 mg/100g), Fe (2.2 mg/100g) and p (53 mg/100g), whereas moisture (88.8%), protein (0.7%), fat (0.5%), carbohydrate (6%), total sugars (5.6%), crude fiber (2.4%), Ca (34 mg/100g), Fe (0.4 mg/100g), p (25 mg/100g), Na (40 mg/100g), K (240 mg/100g), Mg (9 mg/100g), Cu (0.02 mg/100g), Zn (0.2 mg/100g), carotenes (5.33 mg/100g), thiamine (0.04 mg/100g), riboflavin (0.02 mg/100g), niacin (0.2 mg/100g), vitamin C (4 mg/100g) and energy value (126 kJ/100g). The edible portion of carrots contains about 10% carbohydrates having soluble carbohydrates ranging from 6.6 to 7.7g/100g and protein from 0.8 to 1.1g/100g in 4 carrot cultivars reported 1.67–3.35% reducing sugars, 1.02–1.18% non-reducing sugars and 2.71–4.53% total sugars in 6 cultivars of carrot.

AGROTECHNIQUES

It prefers a sunny position and a well-drained neutral to alkaline soil. This species is the parent of the cultivated carrot. Cultivation is mainly done in summer and winter. Different types of varieties are available. Mid. and high sandy loam soil are important for its cultivation. Inorganic and organic fertilizers are applied at the time of land preparation. Proper irrigation is needed. After 3 months of seed sowing, harvesting is done. The plant is harvested in July and dried for later use. Yield is 200-250 kg per hectare.

45. Common name: Dheki, Bhut raj, Maidenhair Creeper
Scientific name: *Lygodium flexuosum*
Family: Lygodiaceae

INTRODUCTION

Lygodium flexuosum is a slender, evergreen, climbing fern with a short-creeping rhizome. All species of the genus have an elongated climbing rachis that has the capacity for indefinite growth, often reaching lengths of several metres. It twines around other plants for support, often climbing up from the shade into a sunny position. *Lygodium flexuosum* is widespread in eastern Asia, adversely affecting rice crops and natural lowland vegetation. It has confirmed medicinal properties. The plant is harvested from the wild for local medicinal use and for its stems used in basketry. The plant is also grown as an ornamental, being valued especially for its ability to quickly cover unsightly structures and providing shade for other plants.

HABITAT

Large terrestrial climbing ferns are common everywhere on low lands, in fairly open places, in teak and bamboo forest, in low country and up to elevations of 1,000 metres. They are generally not found in shady evergreen forest. They can climbs trees and scrambles in open places in deciduous or mixed forests.

MEDICINAL PROPERTIES

L. flexuosum is an important medicinal plant and its medicinal properties have been reported from all parts of the plant. Some of the scholars of Indian System of Medicine reported that the plant may be 'Rudra Jata', an intermediate drug in classical text of Ayurveda. In the treatment of jaundice, the rhizome and root are useful. Kadar tribes of South Western Ghats of India apply the leaf paste all over the body for 7 days to cure jaundice. Fresh root is boiled with mustard oil and is used for massage. Extract of the rhizome is used to cure gonorrhea in India. The ash of plant is used for treating herpes. It is also used to diminish inflammation and wounds; to treat ulcer, various respiratory diseases, general disorders, muscles sprains and had the potential to act as the pain killer. *L. flexuosum* extract had anti-proliferative and apoptotic activity in both cancer cells. N-hexane extracts of the plant are responsible for the possible hepato protective action. Rabha, Oraon and Mech tribes in Jalpaiguri district of West Bengal, India uses the root in jaundice and stomach pain. This fern reported to exhibit anti fertility activity. Lodhas of West Bengal believe *L. flexuosum* to be adobe of gods. During the time of scarcity, Chitwan people use it as vegetable for sustenance they collect tender plant and its parts from the common land. It is used as fodder and forage resources of common land in Western Chitwan and can feed domestic animals to treat foot and mouth diseases.

OTHER USES

The stem is used for tying rice sheaves. Splints prepared from the stems are used in the manufacture of baskets, hats, and fancy boxes. The splints can be combined with buri (*Corypha* spp.) or some other fibre to make an assortment of fancy articles viz., cigarette cases or pocketbooks. The effect is very pleasing, particularly when the plant stem is black. It is used in the manufacturing of basket, hats, bags and other fancy articles.

BOTANY

It is a rhizomatous perennial fern, with climbing rachis has up to 2.5-mm thickness. The dwarfed branches are up to 3-mm long. It is slightly pubsent, substerete, flattened on one side, narrowly winged or tetragonous. Rhizomes are short, creeping, underground, dichotomously branched, young regions clothed by stiff brown hairs and producing a single row of climbing leaves. Primary rachis branches are not elongated. Secondary rachis branches are bearing alternate pinnately arranged leaflets, usually 3 or 4 (occasionally 5 or 6) on each side and a simple or forked terminal leaflet, the whole being 15-30 cm long.

CHEMICAL CONSTITUENTS

L. flexuosum is a rich source alkaloidal constituent it also constitute of flavanoids, saponins and cumarins. A new triterpene ester, an anthraquinone has been found in this plant. Lygodinolide was isolated from the methanolic extract of *L. flexuosum*. It has been found to contain a new compound characterized as O-P-coumaryl dryocrassol, besides drycrassol itself, tectoquinone, kaempferol, kaempferol-3-β-D-glucoside, β-sitosterol and stigmasterol. In Lygodiaceae ferns, antheridiogens were reported to be GA-related compounds. Antheridic acid is a major antheridiogen in four Anemia ferns. The new heterocyclic spiro-terpenoid niveulone which is found to be similar to the compound lygodinolide and niveulone was isolated from the cultural fluid of the ascomycete *Dasyscyphus niveus* and its chemical structure and relative configuration were determined by spectroscopic techniques.

AGROTECHNIQUES

The roots can be planted in a shady position, allowing the plant to climb up into the sun. It prefers a neutral to slightly acid soil. It is mostly propagrated through spores, division and serpentine layering. An actively growing frond is pinned down onto the soil surface at each node, where it may root and shoot out as a new plant.

46. Common name: Parseley
Scientific name: *Petroselinum crispum*
Family: Apiaceae

INTRODUCTION

Parsley has a long history of use. The ancient Greeks believed that it sprang from the blood of Archemorus, the forerunner of death, and so did not eat it but used it for making wreaths to adorn the dead. The Romans wore garlands of it at feasts in the belief that it would prevent intoxication. They kept it away from nursing mothers, however, believing that it could cause epilepsy in the infant.

HABITAT

Anthropogenic (man-made or disturbed habitats), meadows and fields.

Medicinal Properties

Parsley is a commonly grown culinary and medicinal herb that is often used as a domestic medicine. The fresh leaves are highly nutritious and can be considered a natural vitamin and mineral supplement in their own right. It is diuretic and used in treating jaundice, dropsy, cystitis etc. It also acts as a good detoxifier helping the body to get rid of toxins via the urine and, consequently, helps curing diseases like rheumatism. The herb is normally safe at lower doses, but can have toxic effects when used in excess. Parsley should not be advocated for pregnant women as it may provoke a miscarriage by stimulating menstrual flow. However, an infusion of the roots and seeds taken after childbirth encourages lactation and help contract the uterus. All parts of the plant can be used medicinally. The root is the part most often used, though the seeds have a stronger action. Parsley is antidandruff, antispasmodic, aperient, carminative, digestive, diuretic, emmenagogue, expectorant, galactofuge, kidney, stomachic and tonic. Parsley is also a mild laxative and is useful for treating anaemia and convalescents. Unwarranted doses can cause liver and kidney damage, nerve inflammation and gastro-intestinal haemorrhage. A poultice of the leaves has been applied externally to soothe bites and stings. It can also treat cancerous tumours, eye infections, toothache or earache. It is also used against hair loss.

Other Uses

Parsley is frequently used as a garnish or as flavouring in salads and many cooked dishes, but has too strong a flavour to be eaten in quantity for most palates. It is an ingredient of the herb mix 'bouquet garni'. The stems can be dried and ground and used as a food colouring. A tea is made from the fresh or dried leaves. Essential oil obtained from the leaves is used as commercial food flavouring.

The juice is an effective mosquito repellent when it is rubbed into the skin and is also used to relieve the pain of stings and bites. An essential oil obtained from the plant is used in perfumeries for men. An infusion of the leaves is an excellent rinse for dark hair and also helps in the treatment of dandruff.

BOTANY

Three main groups of varieties have developed over a period of time and these have been grouped into subspecies as detailed below:

- *P. crispum crispum*. The curly leafed forms of parsley. This is the more ornamental form, especially when used as a dressing in salads, cooked meals etc. It also has a milder flavour. However the curly leaves tend to hold on to surface water and so they are more likely to contract fungal diseases in the winter.
- *P. crispum neapolitanum* Danert. Italian parsley. This has flat or plain leaves and is considered to have a stronger flavour. This group is also hardier, in part at least because the leaves can shed water easily.
- *P. Crispum tuberosum* (Bernh.)Crov. Hamburg parsley is a very distinct form with a swollen root that is used as a vegetable. The leaves are not of such good quality as the preceding forms, but can still be used as flavouring.

In general, the plant is an erect, copiously branched, biennial to perennial herb, 30-100 cm tall, aromatic in all parts, glabrous. Root system slender, fibrous, taproot up to 1 m long, sometimes thickened, with a radical rosette of leaves when young. Stem is terete, sulcate, hollow. Leaves alternate, 1-3-pinnately compound, dark green, glossy, flat or curled, sheathing at base; petiole longest in lower leaves; pinnate long stalked, leaflets obovate-cuneate to finely linear, divided into acute segments, higher leaves gradually less divided, topmost one consisting of a few acute segments only. Inflorescence a terminal or axillary compound umbel; bracts 1-3-foliolate, rather short; peduncle up to 12 cm long; primary rays 5-20, 1-5 cm long; bracteoles 3-8-foliolate; secondary rays (pedicels) 3-15, 2-5 mm long. Flowers small, yellow-green, bisexual; calyx obscure; corolla consisting of 5 petals, each one suborbicular to obovate, up to 1 mm × 0.5 mm, submarginate with an inflexed apical lobe; stamens 5; pistil with inferior, 2-carpelled ovary, each carpel with a thickened stylopodium, a style and a small globose stigma. Fruit a schizocarp, ovoid, 2-3 mm long, at maturity splitting into 2 mericarps, each one with 5 narrow ribs.

Chemical Constituents

Parsley is very rich in iron, iodine and magnesium. It is also a good source of other minerals and the vitamins A, B and C. The leaves yield about 1% essential oil, whilst about 6% is obtained from the seed.

Agrotechniques

A plant of the temperate zone, it can also be cultivated in the subtropics and tropics. Parsley, particularly the more upright forms, grows convincingly well at low elevations in the tropics. It grows best in areas where annual daytime temperatures are within the range 11 - 20°C. When dormant, the plant can survive temperatures down to about -5°C, but young growth can be severely damaged at 0°C. It prefers a mean annual rainfall in the range 900 - 1,500 mm. It prefers a moist well-drained soil in sun or partial shade. High temperatures can adversely affect growth, so light shade is often helpful. It is tolerant to moderate level of salt in the soil and prefers a pH in the range 6 - 7.5.

Leaves may be harvested 70 - 100 days after transplanting, and harvesting may continue for as long as local climatic conditions allow. The leaves should be harvested regularly in order to encourage fresh growth and get maximum yields. The leaves are difficult to dry but are easily frozen. For drying they require a well-ventilated room that receives long hours of sunlight - the leaves need to be quite crisp if they are to store.

47. Common name: Jayanti, Sesban, Egyptian rattlepod
Scientific name: *Sesbania sesban*
Family: Fabaceae

INTRODUCTION

It is a versatile plant, providing food, medicines, fibre, fuel etc. It is mainly grown for local use and often grown in agroforestry systems to endow with shade, shelter organic matter etc. It is used as a forage (grazed or cut-and-carried), and as green manure. Leaves, flower and seeds are eaten by humans. Sesbania is widely distributed and cultivated throughout semi-arid and sub-humid tropical regions.

Habitat

Plants are common along stream banks, swamp edges and moist and inundated bottomlands from sea levels upto an altitude of 2300 m.

Medicinal Properties

This can be useful to prevent insect bite, cough in children. The fresh roots and leaves are used to treat scorpion stings, boils and abscesses. The leaves are considered to have antibiotic, anthelmintic, antitumor and contraceptive properties. A decoction is used to treat sore throat, gonorrhoea, syphilis, spasmodic fits in children and jaundice during pregnancy. Oil obtained from the seeds is accorded special properties in ayurvedic medicine and is reported to have antibacterial, cardiac depressant and hypoglycaemic actions. From the seeds of the plant, the saponin, stigmasta-galactopyranoside, is isolated, which is believed to be with spermicidal and haemolytic activity.

Other Uses

Flowers cooked, they are included, perhaps as a decorative or festive ingredient, in foods such as omelettes. The leaves and flowers are eaten as vegetable. The seeds are ground into a powder and fermented to make a paste, known as 'soumbara', used as flavouring. The seeds are usually consumed after soaking for 3 days and then cooking, as the fresh seeds are proved to be poisonous to humans. The seeds are also ground into a powder and fermented to make a paste, known as 'soumbara', used as a flavouring. Flowers are generally mixed as a decorative or festive ingredient in foods viz., omelettes. The leaves and flowers are eaten as vegetable.

BOTANY

The plant is an erect softwooded shrub, 10-12 ft in height. The plants cannot give bushy appearance. The leaves can be similar to emli leaves. Sesbania is a short-lived shrub or small tree with a slender crown. Plants with closer spacing usually produce a single stem, but with distance result in many side branches from the base. These branches give the tree a shrubby appearance.

Leaves are deep green, stipules deciduous, leaflets 20-30 in number, membranous, glabrous, linear-oblong, obtuse. Inflorescence is pendulous racemes. Flowers are yellow with brown spots outside, calyx membranous, corolla mottled with brown-purplish spots. Fruits are pod, pendulous, 15-18 cm long, seeds 20-30.

CHEMICAL CONSTITUENTS

The seed contained 4.53% moisture, 2.51% ash, 6.11% crude lipid, 20.22% crude protein, 6.41% crude fibre, 60.51% carbohydrate and food energy value of 377.84 kcal/100g. The pod contained 9.54% moisture, 4.22% ash, 5.41% crude lipid, 8.83% crude protein, 30.34% crude fibre, 41.68% carbohydrate and food energy value of 250.67 Kcal/100g. Antinutritional analysis revealed 0.40% saponin, 4.33mg/100g oxalate, 6.93% alkaloid in seeds while 5.17% saponin, 1.33mg/100g oxalate, 4.91% alkaloid were revealed in the pods. Results from mineral analysis revealed that the seed contained 32.50mg/100g Iron, 59.00mg/100g Manganese, 175.50mg/100g Calcium, 277.20mg/100g Sodium, 2.50mg/100g Cobalt, 0.60mg/100g Nickel, 0.30mg/100g Zinc and Magnesium was not detected. The mineral analysis for the pod depicted 798.83mg/100g

Calcium, 97.76mg/100g Magnesium, 65.79mg/100g Iron, 3.77mg/100g Manganese, 3.05mg/100g Cobalt, 2.52mg/100g Sodium, 0.25mg/100g Nickel and 0.90mg/100g Zinc. The results obtained from vitamin analysis revealed that the seeds and pods of *Sesbania sesban* are excellent sources of B vitamins and vitamin E.

AGROTECHNIQUES

It is a plant of the subtropics and tropics (at higher elevations) generally found at elevations of 100 - 2,300 metres. It grows best in areas where annual daytime temperatures are within the range 18 - 28°C, but can tolerate 10 - 45°C. The plant can survive temperatures down to about -5°C when it remains at dormant stage. It prefers a mean annual rainfall in the range 800 - 2,000 mm. The plant requires a sunny position and can tolerate saline, acidic and alkaline soils and moisture stress to a considerable degree. The plant has an outstanding ability to withstand water logging and is ideally suited to seasonally flooded environments. When flooded, it initiates floating with adventitious roots and protects its stems, roots and nodules with spongy, aerenchyma tissue.

Cutting frequencies are generally 3 - 4 cuts in a year, but up to 8 cuts are possible in some areas. Yields have ranged from 4 to 12 tonnes per hectare dry matter per year, depending on location. Cutting height can also influence yield. A cutting height within a range of 50 - 76 cm favours plant survival and productivity.

48. Common name: Rai sarshe, summer rape, winter rape
Scientific name: *Brassica napus*
Family: Brassicaceae

INTRODUCTION

Brassica napus is annual/biennial growing to 1.2 m (4ft). It is in flower from May to August. The species is basically hermaphrodite (having both male and female flowers) and is pollinated by bees.

HABITAT

Typical habitats include cropland, weedy fields, roadsides, gravelly areas along railroads, and waste areas. This plant is usually found in areas with a history of disturbance where there is scant ground vegetation.

MEDICINAL PROPERTIES

It is regarded as as plants for future. The root is emollient and diuretic. The juice of the roots is used in treating chronic coughs and bronchial catarrh. The seed, powdered with salt is said to be a folk antidote against cancer.

Rape oil is used in massage and oil baths. It is believed to strengthen the skin and keep it cool and healthy. With camphor it is applied as a remedy for rheumatism and stiff joints. It is dropped into the ear to relieve earaches. However, advice from a professional is necessary before using this plant medicinally.

Other Uses

The seed contains up to 45% of edible semi-drying oil. It is potentially used as a luminant, lubricant, ingredients for soap making, etc. Rapeseed oil has prospective market in detergent lubrication oils, emulsifying agents, polyamide fibres, resins, and as a vegetable wax substitute. The most common use for the oil is still in the production or erucic acid, a fatty acid used sequentially for manufacturing other chemicals. The seed husks are used in plastering house walls. A good green manure, the deep taproot improves drainage and loosens heavy soils.

Botany

Brassica napus is an annual or biennial species. The stems are erect, simple to freely branched, glabrous or sparsely hairy and can grow up to 1.5 m tall. Leaves are waxy with a glabrous underside and often have an enlarged base that partially clasps the stem. Inflorescences are racemes which form on the main and axillary branches. Flowering begins at the base of the raceme and the buds form above the open flowers. The pale yellow flowers have 4 sepals and 4 diagonally opposite obovate petals arranged in the form of a cross when viewed from above. The stamens are tetradynamous with four long and two short stamens in each flower. The ovary is superior.

The fruit is a linear cylindrical silique with slight constrictions at regular intervals and dehiscent valves in the lower 4–10 cm segment of the fruit. The seeds are arranged in a single row in the fruit. The upper 3.5–5.0 mm thick segment of the silique is narrow and is usually seedless. There may be 15 or more seeds per silique. Seed formation begins in the lower one-third branches of the main stem. Seeds are translucent and turn green when full sized. At maturity, seeds are spherical, reddish brown, brown or black, 1.8–2.7 mm in diameter with a finely net-veined surface.

Chemical Constituents

Glucosinolates represent a group of phytochemicals found in 15 botanical families of the order of *Capparales* and are very abundant in *Brassicaceae*. Analyses conducted on a collection of 113 varieties of turnip greens (*Brassica rapa* L.), cultivated in two different sites in Spain, showed glucosinolates contents ranging from 12 to 70 μmol/g DW at one site, and from 7 to 60 μmol/g DW at the other site.

Agrotechniques

It can grow in semi-shade (light woodland) or no shade. It likes moist soil and usually prefers full sun and well-drained fertile soil preferably alkaline soil. It may also be grown in heavy soil and cool moist conditions. It can adapt a wide range of soil pH from acidic to basic (alkaline) soils and can tolerate very acid and very alkaline soils also. Sunny days and cool nights are favourable for plant growth whilst dry weather at harvest time is essential.

Major Chemicals Available in Various Non-Descript Leafy Greens

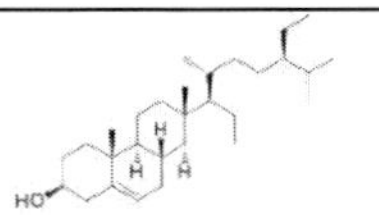 B-sitosterol (*Amorphophallus* sp, ol)	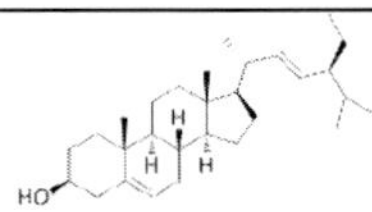 Stigmasterol (*Amorphophallus* sp, ol)	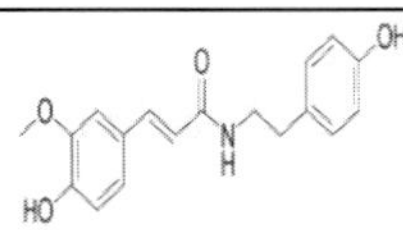 N-cis-Feruloyltyramine (*Ipomea sp, kalmi*)	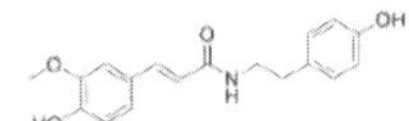 N-trans-feruloyltyramine (*Ipomea sp, kalmi*)
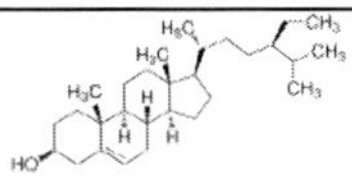 β-sitosterol (Gadalpata)	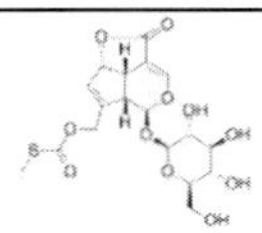 Paederoside (Gadalpata)	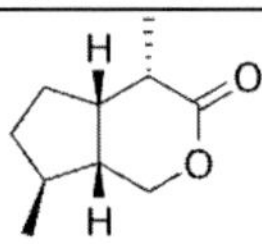 Iridoid (Gadalpata)	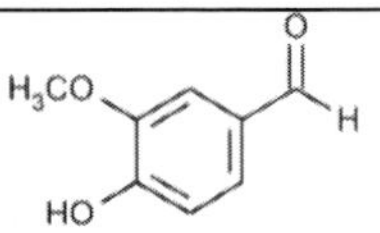 Vanillin (Gima)
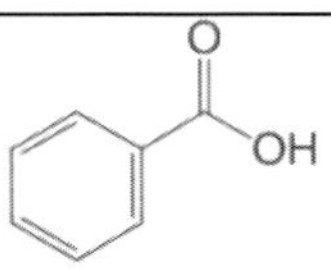 Benzoic acid (*Glinus* sp, Gima)	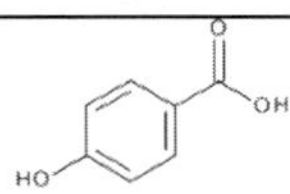 4-hydroxybenzoic acid (*Glinus* sp,Gima)	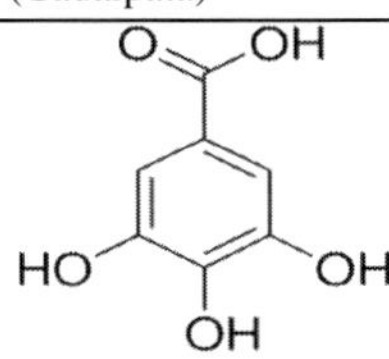 Gallic acid (Chikni)	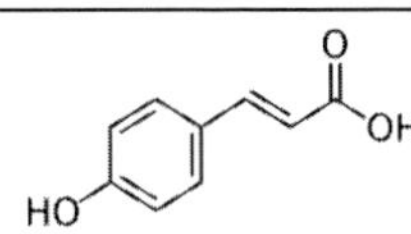 Coumaric acid (*Polygonum* sp,Chikni)
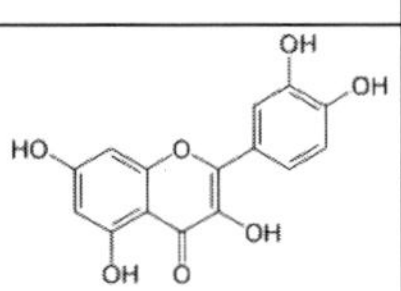 Quercetin (*Polygonum* sp,Chikni)	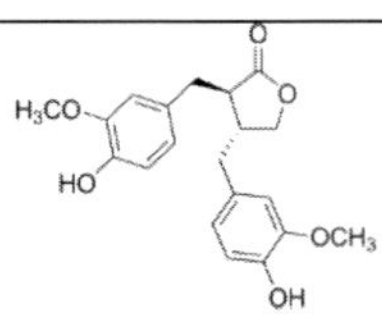 Matairesinol (*Cicer*sp, Chhola)	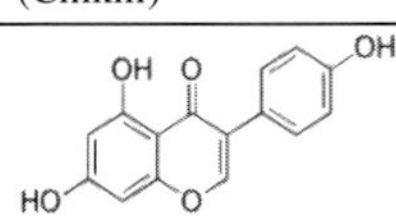 Genistein (*Cicer* sp, Chhola)	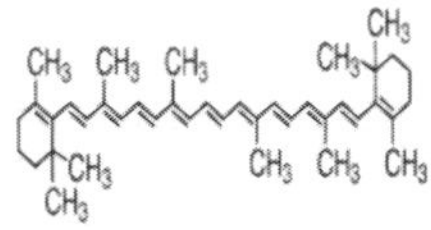 β-carotene (*Cicer* sp, Chhola)
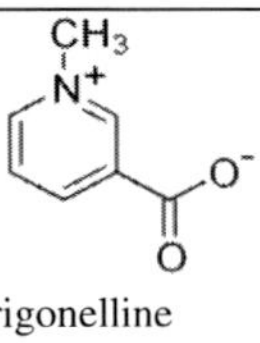 Trigonelline (*Talinum sp*, tokpalong)	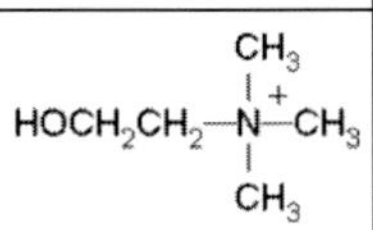 Choline (*Talinum sp*, tokpalong)	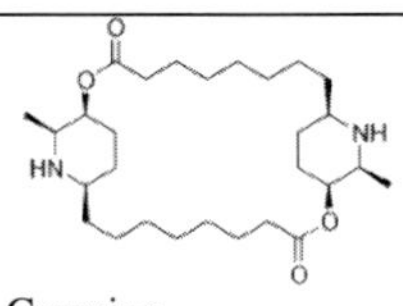 Carpaine (*Talinum sp*, tokpalong)	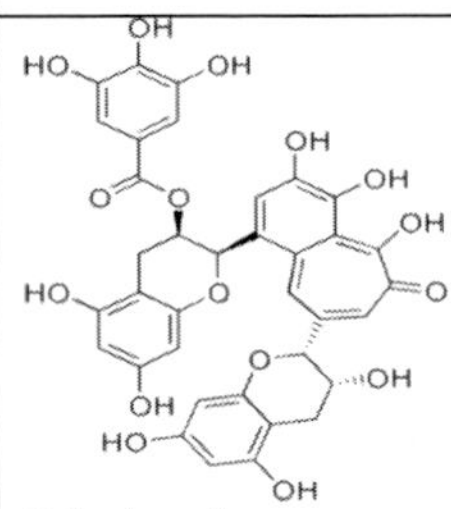 Polyphenols (*Amaranthus* sp, Datanote)

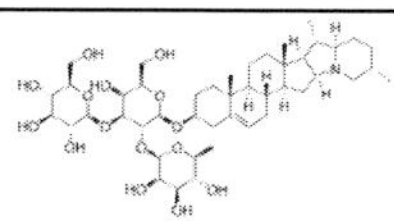 Saponins (*Amaranthus* sp, Datanote)	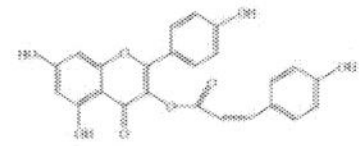 Castilliferol (*Centella* sp, Thankuni)	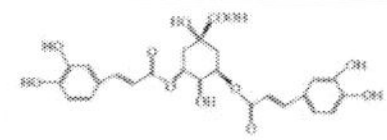 Isochlorogenic acid (*Centella* sp, Thankuni)	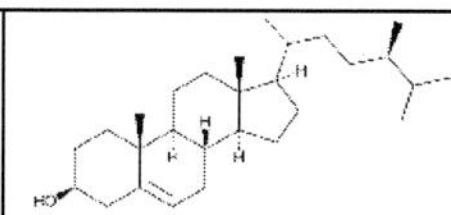 24α- ethylcholest-7-enol (*Trichosanthes* sp, Palta)
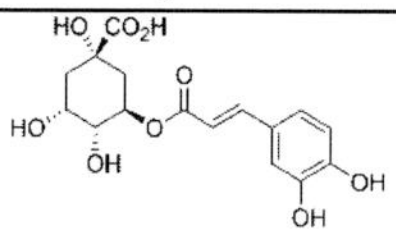 Chlorogenic acid (*Corchorus* sp, Pat)	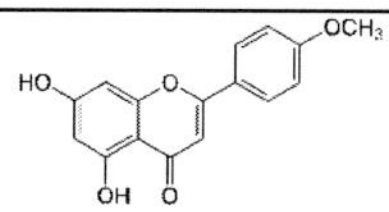 Acacetin (*Basella* sp, Pui)	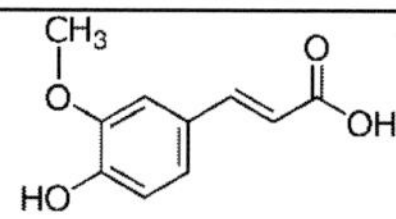 Ferulic acid (*Basella* sp, Pui)	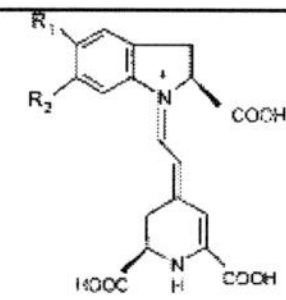 Betacyanins (*Basella* sp, Pui)
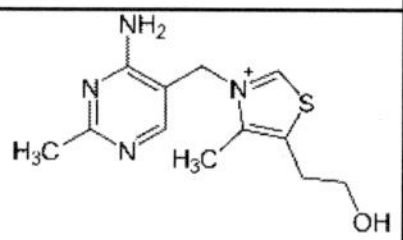 Thiamin (*Spinacia sp,* Palong)	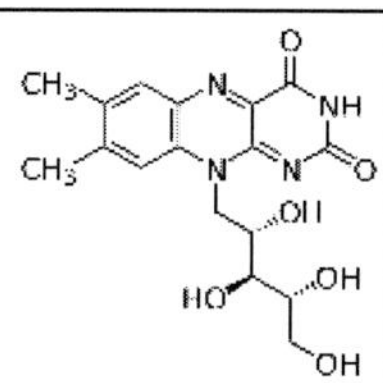 Riboflavin Riboflavin (*Spinacia sp,* Palong)	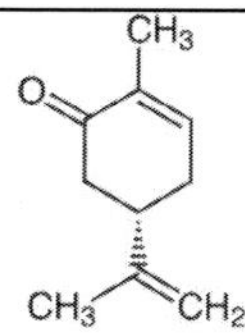 Carvone (*Mentha*sp, Pudina)	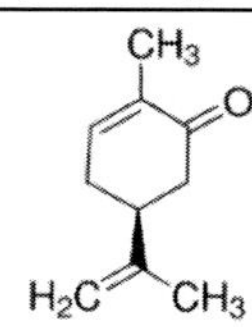 (S)-carvone (*Mentha* sp, Pudina)
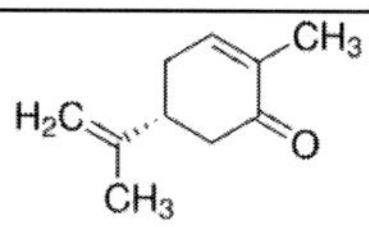 (R)- carvone (*Mentha* sp, Pudina)	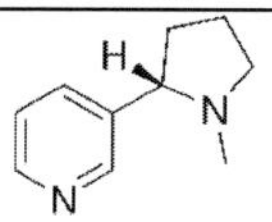 Nicotine (*Bacopa* sp, Brahmi)	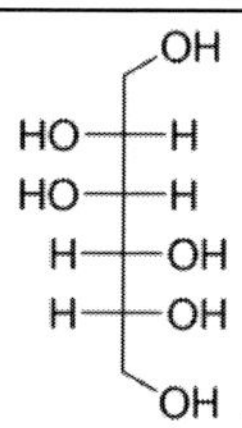 D-mannitol (*Bacopa* sp, Brahmi)	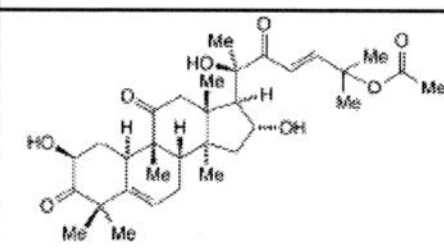 Cucurbitacins (*Bacopa* sp, Brahmi)
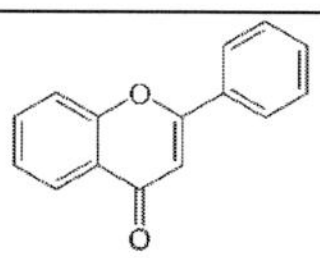 Flavonoids (*Chenopodium* sp, Bathua)	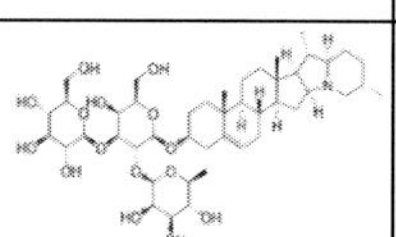 Saponins (*Chenopodium* sp, Bathua)	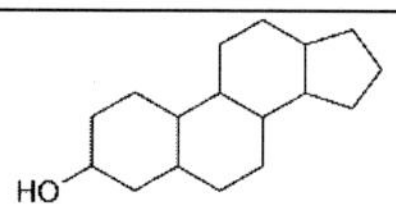 Sterols (*Chenopodium* sp, Bathua)	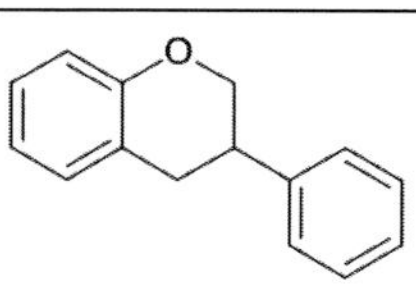 Isoflavanoids (*Sesbania* sp, Bokful)

(Continued)

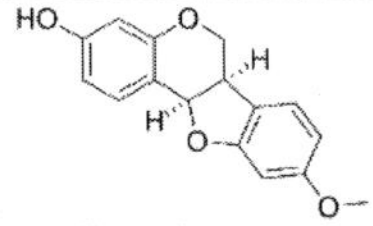 Medicarpin (*Sesbania* sp, Bokful)	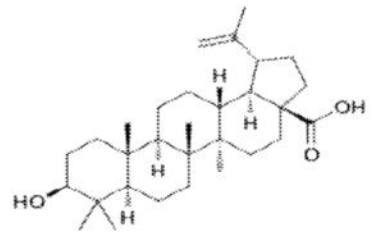 betulinic acid (*Sesbania* sp, Bokful)	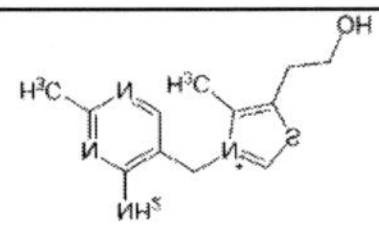 Thiamine (*Sesbania* sp, Bokful)	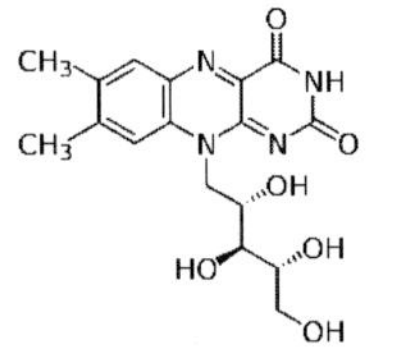 Riboflavin (*Ipomea* sp, Mistialu)
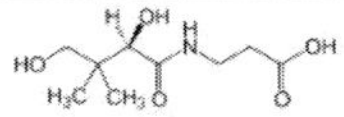 Pantpthenic acid (*Ipomea* sp, Mistialu)	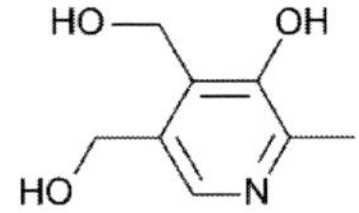 Pyridoxine (*Ipomea* sp, Mistialu)	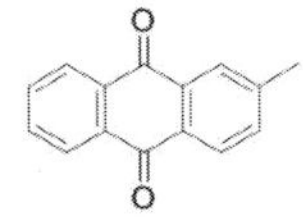 Tectoquinone (Dheki sak)	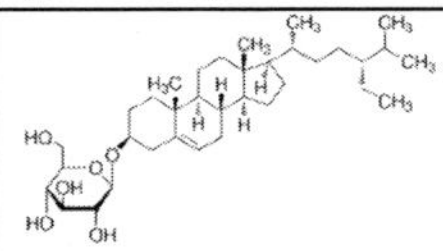 β-Sitosterol (*Lygodium* sp, Dheki)

Chapter 6

EPILOGUE

The sweetest is the music that fiddles on an unknown string, and so does for non-descript agriculture!

The whole world is desperately looking for a vaccine that would save the ailing civilization. Else, we have another way to go winner in this battle by creating comprehensive immunity in our body that can defeats a spectrum of antigens inflicted into our physiology by a virus at any point of time.

We earned green revolutions for so many nations by applying a trio of seed-fertilizer-irrigation techniques, saved the humanity from hunger, and, at the same time, disrupted the mother ecology so ruthlesslely down to stratum which is very difficult to retrieve. At the end of this century, barring Russia, Canada and some few more, the overall productivity of crops would go down, for example, India has to experience a production fall to the tune of 30 per cent in agriculture.

So this is the high time, perhaps the penultimate opportunity to rethink, reengineer and redesign our food ecology and production function.

Million hectares of lands are over exploited with coercive green revolution technology by making the new agriculture regime a docile supplier to industry, market, urbanes,corporate sector, trades, nevertheless, it served the hungry bowls beyond class, and scales.

The non-descript leafy vegetables are one of some few options that can drive us to a valley of new hope by offering us a pool of unexploited genes and ecosystems that are yetsoever happy by not being polluted by wrong agriculture and antropogenic interventions. Through this book, a humble attempt has been endeavoured to elicit the scopes of non-descript agriculture imbued with a re-aligned and re-engineered focus on securing food and immunity for humanity in a new-normal world order. The land remains undisrupted beyond our myopic agriculture and itervention, the genes embeded with pristine ecology dancing with mother ecology since the beginning of human civilization, the knowledge remains in captivity by becoming ignored and overlooked, all are forming a new symphony of gene-land-knowledge to save all the live forms over earth and down the mantle of soil as well. Non-descript vegetables are expected to be loaded with natural immunity treasures, therapeutics and buffers to protect ecological resilience. We can also relate and relegate non-descript characters with the conventional exploitative agriculture to make a mutually complementary benefits of having sufficient food and nutrition for the people and generation at present and in posterity.

The future stand of research on non-descript agriculture are as follows:

i. Idendification, isolation and standardization folk genes that have been integral to an unorthodox ecosystem over centuries ever and creating a gene repository for further study and analysis.
ii. Elucidation of ecolological resilience and the role of non-descript germplasms which would help configuration of a new regime of ecological services.
iii. Immunity research to get a boost from therapeutics explored from non-descript leafy vegetables could be of great stand for herbal medicine and healthcare.
iv. It will be instrumental in overhauling the classics of agro-ecological traditions, rather would be more conceptually invasive and exploratory.
v. The cliches of land can be repaired and upgraded for being innovative and more inclusive by considering rooftops, bogs and

marshy lands, unnoticed terrains, deserts and grasslands, degraded forest regimes, even the roadside leftover lands, experiencing and enriched with plenty's of rains, sunshine, micro flora and fauna, soils like podsols or latosols, sometimes recipients of municipal swearage,both solid and liquid, all are contributors to regime of a non-descript agriculture.

vi. A change in life style may invite the graceful inclusion of non-descript life style altogether! So, ethnic food, drinks and agriculture are to follow an autosocialization process to make life to become an elevate neo-normal pursuits, rather than a classical life process.

vii. A well designed and maneuvered ICT application would be more imminent in online nutrition, hygiene, food and agriculture advocacies well adapted to a non-descript agri-horticultural regime.

viii. A robust platform needs to be created through research, extension and innovation across the globe to make a new age food production and nutritional security system well discernible proposition and well adatable practices by people across the class, scale and echelons of the entire humanity dwelling on our mother earth.

REFERENCES

Austin, J. E.andZeitlin, M.E. (1981). *Nutrition interventions in developing countries: An overview*. Cambridge, Mass., Oelgeschlager, Gunn and Hain, Publ.

Austin, J. E. (1981). *Nutrition programmes in the Third World: Cases and readings*. Cambridge, Mass., Oelgeschlager, Gunn and Hain, Publ.

Beaton, G. H. and Ghassemi, H. (1982). Supplementary feeding programs for young children in developing countries. *Amer. J. Clin. Nutr.,* 35: 864-916.

Carr,A.C. and Maggini, S. (2017). Vitamin C and Immune Function. National Library of Medicine. *National Center for Biotechnology Information*, 3, 1211.

Chakraborty, A. and Acharya, S. K. (2019), Impoverishment and Vulnerability of The Farmers: A Harsh Reality. *Indian Research Journal of Extension Education*, 19, 95-98.

Coates, J. (2013). Build it back better: Deconstructing food security for improved measurement and action. *Global Food Security,* 2, 188-194.

Coates, J., Frongillo, E.A., Rogers, B.L., Webb, P., Wilde, P.E. and Houser, R. (2006). Commonalities in the experience of household food insecurity across cultures: what are measures missing? *Journal of Nutrition*, 136, 1438S-1448S.

Cobo, M. J. (1987). *Study of the Problem of Discrimination against Indigenous Populations.* New York: United Nations Working Group on Indigenous Populations.

Commission on Intellectual Property Rights, (2004). *Integrating IPRs and Development Policy. Report of the Commission on Intellectual Property Rights*, London. p.191

Cynthia Aranow. (2011). Vitamin D and the Immune System; *Journal of Investigative Medicine*; 59, 881-886.

FAO (2005). "The State of Food and Agriculture," *Food and agriculture organization of the united nations, Rome, 2005. Food and Agricultural Organization. Food security statistics.* Available at: http://www.fao.org/3/a-i0100e.pdf.

FAO (2008). "The State of Food and Agriculture," *Food and agriculture organization of the united nations, Rome, 2008.*

FAO (2012). *Proceedings of the International Scientific Symposium on Food and Nutrition Security information: from Valid Measurement to Effective Decision Making*. 17-19 January 2012. Rome. (available at http://www.fao.org/docrep/017/i3244e/i3244e.pdf). FAO. 2012b. Escala.

FAO. (2018). *Biodiversity reveals the treasures all around us.* http://www.fao.org/fao-stories/article/en/c/1133432 (Accessed on 25.11.2020).

FAO, IFAD & WFP (2014). *The State of Food Insecurity in the World 2014. Strengthening the enabling environment for food security and nutrition. Rome.Food and Agricultural Organization. Food security statistics.* Available at: http://www.fao.org/3/a-i0100e.pdf.

Gadgil, M. (2003). *People's Biodiversity Register: Outline of a Methodology Manual.* Paris and Darwin: UNESCO and UNU.

IFCT (2017). Longvah T., Ananthan R., Bhaskarachary K. and Venkaiah, K. *Indian Food Composition Tables.* National Institute of Nutrition (Indian Council of Medical Research), Ministry of Health & Family Welfare, Govt. Of India, pp. 31-111.

IFPRI (2017). *Global Food Policy Report.* https://www.ifpri.org/publication/2017-global-food-policy-report.

Khodadadi, S. 2015 January; Role of herbal medicine in boosting immune system; *Immunopathologia Persa*, 1, e 01.

Linden, E. (1991). Lost Tribes, Lost Knowledge. *Time* (Sept. 23, 1991), 46-56. http://content.time.com/time/magazine/article/0,9171,973872, 00.html(Accessed on 20.10.2020).

Mazumder, M. and Sarkar, A.K. (2019). Ethnobotanical survey of indigenous leafy vegetables consumed in rural areas of Terai-Dooars region of West Bengal, India. *Journal of Threatened Taxa*, 11, 14612-14618.

Singh, H.B. and Arora, R.K. (1978). *Wild edible plants of India. National Bureau of Plant Genetic Resources, New Delhi.* https://www.purdue.edu/hla/sites/famine-foods/ff_reference/singh-arora (Accessed on 25.11.2020).

Suman,R.S.(2008) Adoption pattern of the farmers with regard to improved vegetable production technologies *J. Dairying, Foods & H.S.,* 27 (2): 148 - 150, 2008.www.arccjournals.com/pdf/Dairying/jdfhs-27-2/jdfhs-27-2-015.pdf.

The National Academies Press; 2006, Institute of Medicine; Dietary Reference Intakes: *The essential guide to nutrient requirements;* Washington (DC).

The National Academies Press; 2010, Institute of Medicine; *Dietary Reference Intakes for Calcium and Vitamin D*; Washington (DC).

Thevarajan, I., Nguyen, T.H.O., Koutsakos, M., Druce, J., Caly, L., van de Sandt, C.E., Jia, X., Nicholson, S., Catton, M., Cowie, B. Tong, S.Y.C., Lewin, S.R. and Kedzierska, K. (2020). Breadth of concomitant immune responses prior to patient recovery: a case report of non-severe COVID-19; *Nature Medicine,* 26, 453-455.

UNICEF (2009). *State of the world's children 2009*. New York: UNICEF, 2009.https://www.unicef.org/publications/index_49924.html.

WFP (2015). Purchase for Progress: Reflections on the pilot, February 2015. *WFP - Year in Review 2015.* publications.wfp.org.WFP.org. Retrieved 2015-04-08.

WFP (2019).*Contributions to WFP: Comparative Figures and Five-Year Aggregate Ranking*. United Nations World Food Programme. 2020-04-19. Retrieved 2020-04-21.

WFP (2020) *Contributions to WFP: Comparative Figures and Five-Year Aggregate Ranking*. United Nations World Food Programme. 2020-04-19. Retrieved 2020-04-21.

Wolonciej, M., Milewska, E. And Roszkowska-Jakimiec, W. (2016). Selenium fertilizers for pastures grazed by sheep. Wool and live weight responses to selenium. *Australian J Agri. Res.*, 45, 877-887.

World Bank. (1991). Indigenous People. *The World Bank Operational Manual, Operational Directive*, OD 4.20 September 1991. Washington D.C.: World Bank.

Zalewski, P. D. 1996; Zinc and Immunity; *Journal of Nutritional Immunology,* 4, 39-101

About the Authors

Prof (Dr.) A. B. Sharangi is one of the Professors of eminence in Horticulture and former Head, Department of Plantation, Spices, Medicinal and Aromatic crops, Faculty of Horticulture, Bidhan Chandra Krishi Viswavidyalaya (Agricultural University), India. He is in the profession of teaching for twenty three years. He spent times in Prof. Cousen's lab in Melbourne, Prof Picha's lab in USA, Dr Dobson's lab in UK and continued research on herbs and spices. One of his papers has ranked among the top 25 articles in ScienceDirect.

He has published about 84 research papers in peer-reviewed journal, 56 conference papers, 22 books from reputed publishers including SPRINGER as well as several book chapters published from SPRINGER NATURE, Taylor and Francis, CRC Press, Nova Science Publishers (USA) and scores of popular scientific articles. Presently he is associated with over 50 international and national journals as editor in chief, associate editor, technical editor, editorial board member and active reviewer. Prof Sharangi has visited abroad extensively on academic mission and obtained several international awards viz., ENDEAVOUR Post-doctoral Award (Australia), INSA-RSE Visiting Scientist (UK), FULBRIGHT Visiting Faculty (USA), Achiever's Award (SADHNA), Marquis Who's Who (USA), Man of the Year-2015 (Cambridge, UK),Outstanding Scientist and Higher Education Leadership Award(Venus International Foundation), Bharat Ratna Mother Teresa Gold Medal Award-2020 (GEPRA), etc. He had several invited lectures in UK, USA, Australia, Thailand, Israel and Bangladesh on specific aspects of herbs and spices, chaired several scientific seminar sessions, delivered a couple of key note addresses in seminar, webinar, e-conferences and associated with a number of research projects as Principal and Co-Principal Investigators having academic and empirical implications. He is an elected FELLOW of WB Academy of Science and Technology, (FAScT). He is also the Fellow of International Society for Research and Development (FISRD, UK), the Society of Applied Biotechnology (FSAB), Academy of Environment & Life Sciences (FAELS), International Scientific Research Organization for Science, Engineering and Technology (FISROSET), Scientific Society of Advanced Research & Social Change (FSSARSC), Society of Pharmacognosy and Phytochemistry (FSPP), International Society of Noni Sciences (FISNS). He is an active member of several science academies and societies like NASI, NABS, ISNS, IAHS (Formerly HIS), CWSS including the New York Academy of Science (NYAS),World Academy of Science, Engineering and Technology (WASET), African Forest Forum (AFF), to name a few.

Prof. (Dr.) S. K. Acharya, former Head, Dept. of Agril Extension and Director, Extension Education,Bidhan Chandra Krishi Viswavidyalaya, Mohanpur,WB, born on 6th October, 1960, started his career as Assistant Professor at BCKV in 1988 and has been in teaching, research and extension over 31 years. An erudite teacher and an elegant speaker, Prof. Acharya had been the topper in his MSc (Ag) academic career in Agril. Extension,and is internationally acclaimed for his unique research domain of Social Entropy and Energy Metabolism, Social Ecology and Environmental Sociology, Enterprise Ecology Framework, Farm Stewardships: The Transition from Conservation, Technology Socialization Process, and The Social Ecology of Livelihood, which have got tremendous policy as well as scholastic importance. Research Publication: 208 in National and International Journals. Book Publication: 93 books authored/co-authored, covering topics Participatory Extension Planning and Management; Value addition and quality management, Extension Strategy for Natural Resource Management; Horticulture Management with Quality Issues, Technology Socialization Process; Fish bio-diversity; Income and Livelihood; Enterprise Management and Entrepreneurship Building; Women's participation in Panchayat; Gender Issues in Floriculture; Technology gap in Floriculture; Enterprise Synergy in Agriculture; Disaster Management; Social Metabolism, Farm Energy

Metabolism, Community Perception on Climate Change , Social Entropy and Chaos, Research areas of Publication: Community Forest Management and Role performance by EDC and FPC; Social forestry and community participation (paper published in Indian Forester); Membership of Distinction: Chancellor Nominee on selection committee; Member of Agriculture Commission; Marketing and Extension Sub Committee, Govt. of West Bengal; Expert member WWF projects in BTR (Buxa Tiger Reserve Project); Expert member; DFID(ODI, UK) project on Primary Education (DPEP); REIA, Visited Italy, France, Germany, China, Sri Lanka and Bangladesh, and already selected by ICAR, NAHEP as visiting faculty for Ohio State University, USA. He delivered 30 Key note addresses in national and International Conferences, Chaired 19 Sessions in International conferences, Editors of a score of national and international journals. He has so far been awarded with 16 best paper awards in National and International Conferences and one best thesis award at national level for thesis work under his guidance. He has been the course coordinator for 9 winter/summer schools and delivered 97 high end lectures for the participating faculties from across India. Organizing Secretary of two International Conferences, held in the years, 2016 & 2018, Jointly Organized by Department of Agricultural Extension, BCKV, WB and Krishisanskriti, New Delhi. His Present RG score: 25.33. He has also been honored to be selected as the Convener of Panel (PE-32) entitled The hunger, poverty and silence, of IUAES, University of Manchester, UK, 2013.Co-PI of ICAR and World bank funded project on Conservation Agriculture. So, far he has successfully guided 16 Ph D scholars from BCKV, CU, KU, Vidyasagar University. He is the fellow, ISEE, IARI, New Delhi

Dr. Suddhasuchi Das is working as a Subject Matter Specialist (Horticulture) at Malda Krishi Vigyan Kendra under Uttar Banga Krishi Viswavidyalaya (Agricultural University), India. He has completed his M.Sc. and Ph.D in Horticulture with specialization in Plantation, Spices, Medicinal and Aromatic Crops in the year 2013 and 2018, respectively. He has worked as Project Associate in RKVY Project at Bidhan Chandra Krishi Viswavidyalaya and as a Senior Research Fellow at ICAR-ATARI Kolkata under Farmer FIRST Programme. He has published about 21 research papers in peer-reviewed journal, 30 conference papers, 4 books and 10 book chapters from reputed publishers. He has awarded the best oral presenters in 3 International Conference. He is associated with as an editorial board member of Agriculture & Food e-Newsletter. He is a Life member of Society for Fertilizers and Environment, Kalyani, India.

INDEX

R

S

T

U

V

W

Z